Germana Karla de Lima Carvalho

Efficiency of an Exotic Parasitoid Reared on Different Substrates

Germana Karla de Lima Carvalho

Efficiency of an Exotic Parasitoid Reared on Different Substrates

Evaluation of the parasitism of Fopius arisanus on irradiated and non-irradiated eggs of Ceratitis capitata

Imprint

Any brand names and product names mentioned in this book are subject to trademark, brand or patent protection and are trademarks or registered trademarks of their respective holders. The use of brand names, product names, common names, trade names, product descriptions etc. even without a particular marking in this work is in no way to be construed to mean that such names may be regarded as unrestricted in respect of trademark and brand protection legislation and could thus be used by anyone.

Cover image: www.ingimage.com

This book is a translation from the original published under ISBN 978-613-9-62296-2.

Publisher:
Sciencia Scripts
is a trademark of
Dodo Books Indian Ocean Ltd. and OmniScriptum S.R.L publishing group

120 High Road, East Finchley, London, N2 9ED, United Kingdom
Str. Armeneasca 28/1, office 1, Chisinau MD-2012, Republic of Moldova, Europe
Printed at: see last page
ISBN: 978-620-7-71937-2

SUMMARY

DEDICATORY

To my family, for all their love, dedication and teachings.

ACKNOWLEDGMENTS

To God for his countless blessings.

To Laura, the reason for all my battles and achievements, and to my husband Ricardo, for his companionship, encouragement and support on this journey.

To my grandmother Marli, for taking such good care of my daughter in my many absences, as well as Sonali and Renato.

To my parents, Joao and Iracema, to whom I owe everything, and to my siblings Clàudia, Joao Paulo and Gil, on whom I could always count.

To my advisor Beatriz Jordao for her generosity in passing on her knowledge, for her competence and professionalism.

To Maylen, for her friendship, advice and teaching.

To Biofâbrica Moscamed Brasil (BMB) for allowing the research to be carried out at this institution.

The BMB team: Jéssica, Géssyca, Aline, Meire, Nilton and Edielson, for all their dedication in carrying out the experiments.

To my friends Clésio and Carla, for their sincere friendship and constant presence at all times, encouraging and supporting me on this journey.

To the teaching staff at ITEP, who contributed so much to the process of building knowledge, and to Robson Mascarenhas, for his support with the statistical analyses.

To the Federal Institute of Education, Science and Technology of Sertao Pernambucano (IF Sertao -PE), for giving me the opportunity to do my master's degree.

"Of everything, three things remained:

The certainty that we are always beginning...

The certainty that we must continue...

The certainty that we will be interrupted before we finish...

SO WE SHOULD DO IT:

From interruption to a new path...

From the fall to a dance step...

From fear, a ladder...

From the dream, a bridge...

From the search... an encounter."

Fernando Pessoa

SUMMARY

Fopius arisanus [Sonan, *1932*] (Hymenoptera: Braconidae) is an egg-pupal parasitoid of fruit flies that has been used very successfully in Hawaii. Introduced to Brazil in November 2012 and, in April 2013, at Embrapa Semiàrido, it has been the subject of various studies to adapt it for laboratory rearing under local conditions. The successful use of this species in biological control programs for fruit flies, *Bactrocera dorsalis* and *Ceratitis capitata* (Diptera: Tephritidae) in Hawaii, makes its use in the control of *B. carambolae* in Amapà and *C. capitata in the* San Francisco Valley promising. In order to develop efficient and viable techniques for its mass production, studies were carried out with different doses of X-radiation, applied to *C. capitata* eggs, to determine a dose that would not compromise the development of larvae up to the pupal stage, but which would prevent the emergence of adults. Next, the parasitism of *F. arisanus* on irradiated and non-irradiated eggs was studied. The dose of 15 Gy prevented the flies from emerging. However, the association between irradiated eggs and parasitism showed negative effects, with low egg-pupa recovery. In addition, natural substrates (with natural and artificial egg infestation) and artificial substrates with agar-water blocks were compared for the oviposition of *F. arisanus.* The results have shown a higher rate of parasitism using the natural substrate (guava - *Psidium guajava*) than that obtained with eggs placed on agar-water blocks. There is a need to study more practical methods for exposing *C. capitata* eggs to parasitism, in order to make mass rearing possible in Brazil.

Keywords: parasitoid, Braconidae, biological control, fruit flies.

1   INTRODUCTION

Brazil is the third largest producer of fresh fruit, with an annual volume of approximately 43 million tons. However, despite its high productivity, it only ranks 20[a] in world exports, with a share of only 2.3% in world trade (IBRAF, 2011). This restriction on Brazil's access to international fruit markets is the result of a combination of various factors, including the trade and phytosanitary barriers imposed on Brazilian products (FARIA, 2004).

The widespread entry and acceptance of Brazilian *fresh* produce on the international market requires adaptation to the regulatory trade measures in force, mainly related to the phytosanitary and food safety barriers in place in most developed countries (BUAINAIN; BATALHA, 2007). A limiting factor for the export of Brazilian fruit is the presence of quarantine pests. Several species of fruit flies (Diptera: Tephritidae) fall into this classification and are considered insect pests of recognized economic importance, since their life cycle, especially the larval stage, is closely related to the development of their host fruit (CRESONI-PEREIRA; ZUCOLOTO, 2009).

Among the various species of fruit flies, the most economically representative for Brazil is *Ceratitis capitata* Wiedemann, 1824 (Diptera: Tephritidae) (ALVARENGA, et al., 2007). The great importance of this species stems from the direct and indirect damage it causes. The former is due to the losses caused by the oviposition of the females, hatching and feeding of the larvae inside the fruit, making it unfit for sale and consumption (ALUJA, 1994; FOLLETT; NEVEN, 2006; ALUJA; MANGAN, 2008). Indirect damage stems from the costs of regulatory measures required to market fresh fruit in countries where this pest is considered to be of quarantine importance. The direct and indirect damage accounted for can exceed two billion dollars a year (CARVALHO; NASCIMENTO, 2002). The presence of *C. capitata* therefore represents an obstacle to the export of Brazilian fruit, since importing countries impose severe quarantine restrictions aimed at preventing the introduction of exotic species of fruit flies into their territories, requiring exporting countries to strictly control these pests (CARVALHO et al., 2000).

Considered the most polyphagous of all the members of the Tephritidae family (LIQUIDO et al. 1991), the species *C. capitata* is native to equatorial Africa and was introduced to Brazil in 1901 (IHERING, 1901). Detected in several Brazilian states, its presence is associated with 58 fruit species from 21 families (ZUCCHI, 2001).

In Brazil, the species' preferred hosts belong to the Rutaceae (orange, tangerine and pomelo), Rubiaceae (coffee), Rosaceae (peach, plum, nectarine) and Combretaceae (sun hat) families, which are exotic (MALAVASI, 2009). However, there are more than 89 species of *C. capitata* hosts described in the country (LEA/ESALQ/USP, 2014). This indicates the great ecological and evolutionary plasticity of the species, which adapts quickly to new environments and hosts.

The most widely used method of controlling insect pests is chemical control, as it produces satisfactory results in the short term. However, its continuous and often inappropriate use has favored the emergence and expansion of problems with pest resistance, the emergence of secondary pests and human and environmental poisoning (KOGAN, 1998; PEDIGO; RICE, 2005). In addition, the decrease in the population of natural enemies of pests and the reduction in biodiversity are direct consequences of the environmental imbalance caused by this method.

Considering the problems associated with the use of insecticides, alternative methods to chemical control have been widely encouraged in integrated pest management programs (CARVALHO et al., 2000). Biological control through the use of natural enemies, for example, is a very promising additional tool, as they use the pest as food, reducing its biotic potential (PEDROSA-MACEDO, 1993). The result can be a significant reduction in the use of pesticides, with less impact on non-target organisms and the environment.

Natural enemies include predators, entomopathogens (bacteria, fungi, nematodes and viruses) and parasitoids. The latter are considered the most important and promising in the biological control of fruit flies (CAMPOS, ARAÙJO, 1994). Among these natural enemies, the parasitoids of the Braconidae family stand out due to their specificity when it comes to using tephritids as hosts. For this reason, they are widely used in biological control programs applied in the United States, Mexico and Spain (PARANHOS, 2009).

The species *Fopius* (*Biosteres*) *arisanus* (Sonan, 1932) (Hymenoptera: Braconidae), which belongs to the Opiinae subfamily, is considered to be one of the only endoparasitoids capable of preferentially parasitizing tephritid eggs, although it also has the ability to attack the first larval stages of its hosts (WHARTON; GILSTRAP, 1983). Native to Malaysia, this parasitoid was introduced to Hawaii at the end of the 1940s, showing encouraging results in population control of fruit flies, especially the species *Bactrocera dorsalis* (Hendel), but with some effectiveness on *C. capitata* (ROUSSE et al., 2005).

Despite its great potential, *F. arisanus* has not been widely used as a biological control agent for tephritids in other parts of the world. This is partly due to the scant information available on its biology, which made it difficult to breed it. However, with the development of techniques that make it possible to breed this parasitoid in laboratories (BAUTISTA et al., 1999), there has been considerable progress in research into morphology and behavioral biology (ROUSSE et al., 2005).

Introduced into Brazil in November 2012 by the "Costa Lima" Quarantine Laboratory (LQCL) of Embrapa Meio Ambiente (CNPMA), *F. arisanus* has been multiplied on *C. capitata* eggs. *capitata eggs* and, since then, several studies have been conducted with the aim of obtaining more information on its biology and behavior, as well as verifying its viability in controlling fruit flies present in Brazil, particularly the species *C. capitata* and *Bactrocera carambolae* Drew & Hancock, 1994, restricted to the states of Amapà and Roraima (ZUCH, 2000; BRASIL, 2010).

In addition to biological control using parasitoids, the association with the sterile insect technique (SIT) increases the efficiency of fruit fly control (CARVALHO, 2006). According to Benedict and Robinson (2003), TIE is the most specific of all the techniques, as it targets a particular species and is compatible with other control methods. TIE is a control method that consists of rearing, sterilizing and releasing, on a large scale, the pest insect you want to control. The sterile insects copulate with the wild ones without generating offspring, resulting in a reduction in the pest's native population (KNIPLING, 1955; MORRISON et al., 2009).

Ionizing radiation is widely used to sterilize insects in programs to suppress or eradicate fruit flies and other pests such as the earwig fly (BAKRI et al., 2005). In entomology, radiation can also be applied to the mass production of parasitoids (WALDER, 2000).

The aim of this study was to evaluate the development of the parasitoid *F. arisanus* on eggs from two strains of *C. capitata, the* bisexual strain and the *tsl* Vienna 8 mutant strain, in order to address the following topics: i) Evaluate the parasitoid's preference for different oviposition substrates; ii) Study the feasibility of using two strains of *C. capitata, the bisexual and the tsl Vienna 8 mutant* strains, as hosts. *capitata*, bisexual and mutant *tsl* Vienna 8, as hosts for *F. arisanus*; iii) Determine the dose of X-rays applied to *C. capitata* eggs to prevent the emergence of fruit fly adults and iv) Evaluate the parasitism of *F. arisanus* on irradiated and non-irradiated eggs;

2 THEORETICAL FRAMEWORK

2.1 Fruit flies

The Tephritidae family, one of the largest in the Diptera order, includes the species considered to be the true fruit flies (CRESONI-PEREIRA; ZUCOLOTO, 2009). They have a wide geographical distribution and their representatives comprise 4,448 species grouped into 484 genera (NORRBOM, 2004). They are phytophagous and, in their larval stage, feed on plant tissue, especially fruit pulp (MALAVASI, 2009). Because of their feeding habits, members of this family are considered to be the most economically significant pests in fruit growing worldwide. In Brazil, seven species of the genus *Anastrepha* Schiner (1868) stand out [*A. grandis* (Macquart, 1846), A. *fraterculus* (Wiedemann, 1830), A. *obliqua* (Macquart, 1835), *A. pseudoparalela* (Loew, 1873), *A. sororcula* Zucchi 1979, A. striata Zucchi 1979, *A. striata Schiner 1868 and A. sororcula* Zucchi 1979]. *striata* Schiner 1868 and *A. zenildae* Zucchi 1979 (ZUCCHI, 2000)] and a single species of the genera *Ceratitis* MacLeay (1829) and *Bactrocera* Macquart (1835), *C. capitata* Wiedemann (1824) *and B. carambolae* Drew and Hancock (1994), respectively.

The population fluctuation of the species is directly dependent on the availability of the hosts it is able to use as food and on the climatic factors of the region. A clear example of this is the population levels of *C. capitata*, which are high in orchards with great diversity and availability of ripe fruit (PARANHOS et al., 2008). The choice of host for oviposition is extremely important in the life cycle of fruit flies, since the development of immature stages, as well as the survival and fecundity of adults depend on the host selected (MALAVASI, 2009).

The life cycle of *C. capitata* can last 21 days at temperatures between 26° and 29°C and comprises four stages of development: egg, larva, pupa and adult (CARVALHO et al., 2000).

The eggs of *C. capitata* are white, translucent, elongated and banana-shaped, measuring around 1 mm in length. The incubation period varies from 2 to 4 days and after hatching the larva feeds on the endocarp where it goes through 3 instars of development, lasting an average of 6 to 11 days. After this period it leaves the fruit to form a puparium in the soil (PARANHOS, 2005).

The fully developed larva measures around 8 mm in length, is àpoda, yellowish-white in color, vermiform in shape, tapered at the front and truncated and rounded at the back (GALLO et al., 2002).

The pupa measures around 5 mm in length, is brownish-brown in color, barrel-shaped and coarcted, i.e. surrounded by the exuvia of the last instar, therefore without any appendages of the future insect visible. Around 9 to 11 days after pupation, the adult emerges (SOUZA FILHO, et al., 2004).

The adult *C. capitata* measures around 4 to 5 mm in length and has a wingspan of 10 to 12 mm. It is predominantly dark yellow in color, with violet-brown eyes, a black thorax on the upper side with symmetrical white designs, and a dark yellow abdomen with two greyish yellow transverse stripes (GALO et al., 2002). In addition to the presence of the ovipositor in the females, sexual dimorphism is characterized by the presence, in the male, of a pair of spatulate bristles in the anterior fronto-orbital region (HUNTER et al., 2002).

To establish a colony of the bisexual strain of *C. capitata*, wild insects of the species are collected in the field and kept in captivity for many generations, under conditions that differ drastically from those found in the wild. This strain has the same characteristics as the wild insects found in the field, but is adapted to artificial laboratory conditions.

The *tsl strain* of *C. capitata* is the strain most commonly used by sterile *C. capitata* biofactories around the world (ROBINSON et al., 1999). This strain belongs to a genetic sexing group, *GSS* (*Genetic Sexing Strain*), whose mutations include *tsl* (*temperature sensitive lethal*) (FRANZ et al., 1996) and *wp* (*white pupae*) (ROSSLER, 1979a). This lineage allows males and females to be separated in their embryonic stages and is based on a reciprocal translocation between the chromosome that determines the male sex and the autosome that carries the marker or mutation used to construct the sexing system (ROSSLER, 1979b; KERREMANS; FRANS, 1994). The sexes are separated early on by exposing the eggs to a temperature of 34°C for a period of 24 hours, which is necessary to kill the female eggs (FISHER; CACERES, 2002; CACERES et al., 2000).

2.2 Sterile Insect Technique (SIT)

Considered a type of autocidal or genetic control, in which the pest is used to control itself (WALDER, 2000), the technique is considered the main tool in many *C. capitata* integrated control programs around the world (DYCK et al., 2005).

TIE consists of large-scale rearing of the insect pest you want to control, sterilizing the insects using irradiation and then releasing these sterilized insects into the field on a weekly basis. Sterile insects copulate with wild ones, but will not produce offspring (KNIPLING, 1955).

In order to sterilize insects, gamma radiation is the most commonly used worldwide, which can come from Cobalt 60 (Co^{60}) or Cesium (Cs^{137}). However, its use poses biological risks, requiring special safety precautions. As a result, the use of these sources for this purpose has been discontinued and the alternatives of special radiation sources (X-rays and accelerated electrons) represent lower operating risks (WALDER, 2000).

Paranhos et al. (2008), comment that the basic premises for the use of TIE in insect control are: sexual reproduction through copulation, females copulating preferably only once and ease of rearing the pest on an industrial scale, on an artificial diet. The efficiency of the technique can also be greater when only males are released into the field, as the likelihood of them copulating only with wild females increases, with reductions in production and release costs.

In order to make it possible to release only males in the field, a mutant strain was developed in the 1980s, whose females emerge from white pupae, so that the white pupae can be discarded and the brown pupae kept for the release of sterile males (CACERES, 2002). A major breakthrough in this respect came with a mutant whose females have lethal temperature sensitivity (tsl) above 34°C, even in the embryo stage (ROBINSON et al., 1999; CACERES, 2002). Therefore, when the aim is to release sterile males in the field, the eggs are treated in a bain marie at 34°C for 24 hours, exterminating all the eggs that would give rise to females.

Câceres et al. (2007) point out that the introduction of this technique in pest control has contributed to the development and creation of new areas in entomology, such as the rearing of insects in artificial media (mass rearing), ecology and population simulation, quality control and radioentomology. The use of sterile insects to control or eradicate an insect population in the 1940s was a revolutionary initiative in modern entomology (DYCK et al., 2005).

The first experiments with X-rays, initially called Roentgen rays, date back to the beginning of the 20th century. In entomology, their application has almost always been limited to a small number of species or for very specific purposes. X-rays are emitted when electrons accelerated by high voltage are thrown against atoms and undergo braking, losing energy. They are more penetrating than cathode rays and are not deflected by the magnetic field. They can pass through different materials of various thicknesses and even liquids. This type of radiation, called ionizing, causes changes in the water molecules of living beings (OKUNO; YOSHIMURA, 2010).

One of the modern applications of X-rays in entomology is the production of sterile insects for fruit fly control. At the end of 2010, the social organization Biofâbrica Moscamed Brasil

(BMB), located in Juazeiro, BA, received its first X-ray machine. With the support of the International Atomic Energy Agency (IAEA), Brazil now has the first biofactory in the world to irradiate *C. capitata* pupae with X-rays to replace the traditional gammacell irradiators based on Cobalt 60.

The biological effects of X-rays are similar to those of gamma radiation, since both are electromagnetic radiation. The advantages of X-rays are: no radiative decay, less shielding, easy transportation and easier application (WALDER; MASTRANGELO, 2011).

2.3 Biological Control of Fruit Flies

Biological control was defined by DeBach (1968) as the action of parasitoids, predators and pathogens in maintaining the density of another organism at a lower level than would normally occur in the absence of those agents.

Biological pest control, together with other conservationist control tactics, such as the sustainable management of natural resources and the preservation of biodiversity, contributes to the sustainability of Brazilian agribusiness (STEFANELO, 2002). It is an important component of the integrated pest management strategy, and according to Parra et al. (2002), it is becoming increasingly important in integrated pest management (IPM) programs, especially at a time when there is a lot of discussion about integrated production towards sustainable agriculture.

Among the natural enemies used in biological pest control, parasitoids are of great economic value because they efficiently regulate the population of pest insects. These organisms use only one host to complete their development cycle. The adults are free-living, feeding on pollen and nectar. These characteristics, combined with the efficiency of the results achieved, support the use of this group of insects in the biological control of herbivore populations in various agricultural crops (GARCIA, 1991).

Most of these insects belong to the order Hymenoptera, one of the four largest, with around 115,000 described species (LASALLE; GAULD, 1993), although it is believed that the total number is five to 10 times greater than that already catalogued. This order is very important, both biologically and economically, but it is still poorly described. It is known, however, that most species are found in tropical and subtropical countries such as Brazil, Ecuador, Peru, Colombia, Mexico, Indonesia, China, Zaire and Madagascar. Among the natural enemies of fruit flies, the parasitoids of the Braconidae family stand out, being widely used in biological control programs applied in the United States, Mexico and Spain (PARANHOS, 2009). They are holometabolous insects, i.e. they undergo complete metamorphosis. The females oviposit on or inside their hosts at any stage of their development. Because of this, these natural enemies can be classified as egg, larva, pupa and adult parasitoids (COSTA et al., 2006).

Some egg parasitoids kill the host even before it begins its attack on the plant. However, this is not a rule, as some parasitize the eggs, but complete their development in the pupal stage of the host. In all cases, whether they are egg parasitoids, larvae, pupae or adults, they reduce the size of host populations in subsequent generations.

The use of biological control agents results in a considerable reduction in the use of chemical products and in the levels of toxic residues on the fruit, thus increasing the quality of the end product and competitiveness in the *fresh* fruit market (ALVARENGA et al., 2006).

It is not uncommon for organisms used in biological control projects and programs to originate from other regions or countries. The transportation of these organisms involves risks (SA, 2001). The voluntary transit of organisms is controlled by the legislation of each country. According to Moraes et al. (1996), Brazil has had legislation for many years on the introduction of exotic species, the shipment of natives abroad and the collection of living organisms by Brazilians and foreigners in national territory.

The transit of biological material for research purposes is carried out through quarantine laboratories, which aim to reduce the likelihood of undesirable organisms entering the country (SA, 2001).

Ordinance 106 of November 14, 1991, issued by the Secretariat of Agricultural Defense of the Ministry of Agriculture, Livestock and Supply (MAPA), established the accreditation of the Quarantine Laboratory for Useful Organisms for biological pest control, located at Embrapa Meio Ambiente (CNPMA), of the Brazilian Agricultural Research Corporation (EMBRAPA), in the city of Jaguariùna, São Paulo (BRASIL, 1991). Following this standardization, the laboratory located at Embrapa Meio Ambiente was named the "Costa Lima" Quarantine Laboratory (LQCL) (SA, 2001).

The work carried out at the LQCL follows quarantine standards and procedures for the exchange of live organisms for research into the biological control of pests, diseases, weeds and other scientific purposes (EMBRAPA, 1995). These standards were approved by the Ministry of Agriculture and Supply (MAPA) by means of Ordinance 74 of March 1994 and Normative Instruction No. 1 of December 15, 1998, also from MAPA, which contains the form for submitting requests for organisms, their respective evaluations, as well as the general characteristics of the organisms introduced (SA, 2001).

Quarantine activities therefore have a conservationist focus, aimed at preventing possible negative environmental impacts from the introduction of an exotic organism.

2.4 *Fopius arisanus* (Sonan) (Hymenoptera: Braconidae)

The species *Fopius* (*Biosteres*) *arisanus* (Sonan) (Hymenoptera: Braconidae) belongs to the subfamily Opiinae, which comprises 26 genera and more than 800 species, most of which belong to the genus *Opius* Wesmael. *F. arisanus* is considered to be one of the only endoparasitoids capable of preferentially parasitizing tephritid eggs (WHARTON; GILSTRAP, 1983; HARRIS & BAUTISTA, 1996), although it also has the ability to attack the first larval stage of its hosts (BAUTISTA et al., 1998), 1998), emerging 18 to 20 days after oviposition, around two days after the emergence of its non-parasitized hosts (ZENIL et al., 2004, MANOUKIS, et al., 2010).

It has a high potential for use as a biological control agent for tephritids. This is due to the fact that, by parasitizing its hosts at early stages (egg stage), *F. arisanus has an* advantage over potential competitors that exploit larval and pupal stages in subsequent parasitism (ROUSSE et al., 2005). In addition, as host eggs are immobile and usually deposited in groups, these structures have no escape mechanism, making them extremely vulnerable to attack by egg-pupa parasitoids (MILLS, 1994). Both arguments may explain the prevalence of *F. arisanus* over parasitoids that exhibit parasitism strategies other than egg-pupation.

Another comparative advantage of the species is its high capacity to flex its ovipositor, making the female parasitoid capable of parasitizing several eggs located close to its oviposition site, without having to change its original position (MONTOYA et al., 2009). In addition, these parasitoids are compatible with organic insecticides. This particularity can make the biological control strategy an important component of Integrated Pest Management (IPM) (VARGAS et al., 2001).

Despite its potential effectiveness in the biological control of fruit flies, *F. arisanus* has rarely been used in other countries for this purpose. This is probably due to the fact that, until the end of the 1990s, little information was available on its biology, which meant that it was difficult to breed it en masse. However, since the development of viable techniques for rearing this parasitoid in the laboratory (BAUTISTA et al., 1999), there has been a considerable increase in research into its morphology and behavioral biology, as well as its potential for use in the biological control of fruit flies (ROUSSE et al., 2005).

Although it is considered a polyphagous parasitoid, to date all the hosts recorded for *F. arisanus* belong to the family Tephritidae and, more specifically, to the genus *Bactrocera* (ROUSSE et al., 2005). Among the host species are *Anastrepha ludens* (Loew), *A. serpentina* (Wiedemann), *A. striata* Schiner, A. *suspensa* (Loew), *B. dorsalis, B. barringtoniae* (Tryon), B. *cucuminata, B. carambolae, B. jarvisi* (Tryon), *B. kraussi* (Hardy), *B. latifrons* (Hendel), B. *neohumeralis* (Hardy), B. *oleae* (Gmelin), B. *papayae* Drew & Handcock, B. *passiflore* Froggatt, B. *tryoni (*Frogatt), *Carpomya vesuviana* Costa and *C. capitata* (QUIMIO; WALTER, 2001, ROUSSE et al., 2005).

Detailed knowledge of the insect's biology is fundamental to the success of any biological control program. Studies detailing the developmental stages of *F. arisanus* are essential for establishing adequate quality control in mass rearing programs (ROCHA et al., 2004). Aspects such as the size and duration of the immature stages of parasitoids vary with the age, size and quality of the host on which they are reared (LAWRENCE et al., 1976; LAWRENCE, 1990).

Previous descriptions indicated that *F. arisanus* had four stages of larval development (IBRAHM et al. 1992). However, based on direct observations of sequentially dissected samples and the morphology of *F. arisanus larvae* in *B. dorsalis* Rocha et al. (2004) believe that this parasitoid has three larval stages and a pre-pupal period. Since the evaluation of the teguments of the early and late third stages did not detect any clear morphological differences.

The parasitoid's eggs are deposited inside the host, more precisely in its embryo, and hatch around 28 hours after laying. The attacked host develops normally and *F. arisanus* remains in its first larval stage until its host reaches the pre-pupal stage. At this point, the parasitoid's larvae moult for the first time and, from then on, they develop, consuming their host inside the puparium until they reach full development and subsequent emergence (ROUSSE et al., 2005).

The total egg/pupa cycle can last 21 days when the temperature varies between 26°C and 29°C (ROCHA et al., 2004). As a result of this parasitism, an adult parasitoid emerges instead of a fly. The emergence of males of *F. arisanus* precedes the emergence of females by one or two days, a phenomenon called protrandia (WANG; MESSING, 2003).

Mating begins a few days after the emergence of the male parasitoids (QUIMIO; WALTER, 2001). The eggs of the females mature quickly, reaching a peak 4-6 days after the emergence of the insect (RAMADÂ et al., 1992). As in all braconids, the females of *F. arisanus* show arrenotocous parthenogenesis, i.e. the males come from the eggs of the females (haploid), without copulation occurring.

In the choice of hosts, Vargas et al. (1991) reported that the physical appearance of the fruit, which hosts the fruit fly, is attractive for *F. arisanus* females to lay eggs. Another possible important sign for attracting parasitoids is the presence of females of the host fly on the fruit (HARAMOTO, 1953). Fruit characteristics such as size, shape and stage of ripeness can influence parasitism by *F. arisanus* (VARGAS et al., 1991; HARRIS; BAUTISTA, 1996).

According to Wang and Messing (2003), *F. arisanus* appears to have an almost perfect ability to discriminate the host, with very little risk of wasting time and eggs in its foraging behavior. Observations showed that the parasitoid spent more time foraging on unexplored fruit than on fruit previously visited by another female of the same species. The average number of eggs laid followed the same trend, and was significantly higher on previously unexplored fruit.

Bautista et al. (1999), reports that offering a large number of eggs from the host is the preferred method for mass rearing of *F. arisanus,* although the percentage of parasitism is relatively modest, around 14 to 21%.

In addition to the quality of the host, environmental conditions are one of the main factors favoring the successful multiplication of a parasitoid. To ensure that mating occurs, parasitoids must be provided with specific environmental conditions such as well-lit and ventilated cages. Lack of air circulation favors a pheromone-saturated environment that discourages copulation (JANG et al., 2010).

In the mass rearing of the species, it is important to obtain a higher proportion of females in relation to males, since they are the "active ingredient" in biological control programs by parasitoids. Fertilized females will generate offspring made up of females (diploids) and males (haploids), which will then be able to establish themselves in the field (PARANHOS; BARBOSA, 2005).

3 MULTIPLICATION OF *F. arisanus* (Sonan, 1932) (Hymenoptera: Braconidae) ON EGGS OF *C. capitata* WIEDEMANN, 1824 (Diptera: Tephritidae), BISEXUAL AND MUTANT LINES *tsl* VIENNA 8.

3.1 Introduction

The species, *F. arisanus,* was imported from the USDA/PBARC, Hilo, Hawaii, and introduced into Brazil in November 2012, through Embrapa Meio Ambiente's "Costa Lima" Quarantine Laboratory (LQCL), with the aim of biologically controlling the carambola fly, *Bactrocera carambolae,* in Amapâ. In April 2013, the first parasitized pupae were sent to the Entomology laboratory at Embrapa Semiàrido, in Petrolina, PE. After multiplication, this institution gave part of the parasitized pupae to Biofâbrica Moscamed Brasil (BMB), in order to maintain a safe colony and, above all, to make it possible to carry out studies involving the use of X-radiation.

The interest in carrying out new studies with *F. arisanus* is based on the characteristics that make this parasitoid an effective biological control agent for fruit flies, being one of the only species of opiine endoparasitoids known to attack fruit fly eggs (WHARTON, GILSTRAP, 1983).

In natural environments, the hosts of most parasitoids are distributed in discrete clusters ("Patches"). Thus, according to optimal foraging theory, the main problem faced by female parasitoids during foraging is the distribution of time in host clusters and the way in which they are exploited, in order to maximize their reproductive success (LOPES et al., 2008), which is very well done by this parasitoid species.

The reproductive and developmental biology of *F. arisanus* on various species of fruit flies has been well documented (BAUTISTA et al., 1998; ZENIL et al., 2004; MONTOYA et al., 2009), as well as aspects related to the location of its host (WANG & MESSING, 2003; ROUSSE et al., 2007). Several authors have described the steps taken by *F. arisanus* in the process of locating a host and oviposition and suggest that the main factor leading to its success is its high foraging efficiency [RAMADA et al., (1992, 1994); BAUTISTA et al., (1998); WANG & MESSING, 2003)].

The adaptation of the parasitoid to the alternative host in which it is to be mass-reared is an important stage in achieving success in biological control programs. To this end, evaluations are needed which should mainly involve the suitability of the alternative host and the biotic and abiotic factors related to the artificial methods of multiplication.

Variations found in parasitism may be related to the rearing technique used, such as: the environmental conditions of temperature, relative humidity and photoperiod, type of host used, oviposition substrate, larval diet, among others (PRATISSOLI; OLIVEIRA, 1999).

Despite the wealth of information on the biology and behavior of *F. arisanus*, previous attempts to multiply it in the laboratory were unsuccessful (HARAMOTO, 1953; CHONG, 1962; SNOWBALL et. al., 1962; RAMADAN et. al. 1992,   1994; WANG & MESSING, 2003). According to these authors, when wild parasitoids were taken to the laboratory, copulation rarely occurred, so the number of females was extremely low, which made it impossible to mass rear the species.

Bautista et. al. (1999) described advances in the mass rearing of *F. arisanus* that made it possible, for the first time, to produce this species of parasitoid on a large scale. In a period of 13 months, almost 10 million parasitoids were produced on *Bactrocera dorsalis* Hendel. In the research carried out by the authors, two important advances were made: the establishment of a strain of *F. arisanus* adapted to artificial laboratory conditions and evidence that the parasitoid's females have adapted to an artificial method of exposing the host.

The fundamental objective of any insect breeding program is to guarantee the production of high quality insects at low costs (FINNEY, FISHER, 1964; KNIPLING, 1966), and one of the important stages in this process is the definition of the best host.

With a view to large-scale production of *F. arisanus* for use in biological control programs in Brazil, this study aimed to: i) Adapt and develop methodologies for artificial rearing of this parasitoid under local conditions in Brazil; ii) Evaluate the parasitism of *F. arisanus on C. capitata eggs on different substrates*; iii) Determine the best strain of C. capitata, bisexual or mutant tsl Vienna 8, on different substrates; iv) Determine the best strain of C. capitata, bisexual or mutant tsl Vienna 8, on different substrates. *arisanus on C. capitata* eggs on different oviposition substrates; iii) Determine the best strain of *C. capitata*, bisexual or mutant *tsl* Vienna 8, to be used as an alternative host in the mass rearing of *F. arisanus*;

3.2 Material and Methods

The work was carried out at the Entomology Laboratories of Embrapa Semiàrido and the Biofâbrica Moscamed Brasil (BMB). The rearing of the insects and the experiments were carried out in rooms with controlled environmental conditions (24 ± 2°C, RH of 50 ± 10%

and photophase of 14 h).

Host breeding - *C. capitata*

For the multiplication of *F. arisanus*, as well as for the development of the research, the bisexual and mutant *tsl* Vienna 8 strains of *C. capitata were* used, both from colonies kept in the laboratory and multiplied on an artificial diet developed by Damasceno (2013). An adult cage (30 x 30 x 30 cm) with an aluminum frame, acrylic base and voal fabric sides was set up for each of the strains. 100 mL of pupae were placed in each cage, corresponding to approximately 4,200 adults (1 ♀:1 ♂), a container with water and another containing an adult diet (3:1, sugar: hydrolyzed protein -Biones®) (SILVA NETO et al., 2012). Each cage was placed on a tray containing filtered water for egg collection.

Rearing the parasitoid *F. arisanus* on the bisexual strain of *C. capitata*

Pupae parasitized by *F. arisanus* were placed in cages (30 x 30 x 30 cm) with aluminum edges, fixed acrylic bottom, top and sides with fiberglass mesh (1 mm mesh), containing water and bee honey "ad libitum".

For multiplication, guavas (*Psidium guajava* L.*)* at the beginning of ripening were used and subjected to natural infestation by *C. capitata by* introducing them into laboratory cages, where they were exposed to oviposition for one hour. They were then offered to *F. arisanus* adults, from the 9th day of age, for 24 hours. After parasitization, the guavas were placed in plastic jars (500 mL) containing a 2 cm layer of vermiculite and sealed with a cloth until the pupae were formed.

The duration of the egg-adult cycle, parasitism rate (P), pupal viability (PV) and sex ratio (SR) were evaluated using the following formulas:

P = [number of parasitoids/ (number of flies + number of parasitoids)] X 100

VP= (number of emerged adults / total pupae) X 100

RS= number of females/ (n^o . females + n^o . males)

Comparison of methods for exposing *C. capitata* eggs *to the* parasitoid *F. arisanus*

Two methods of exposing the host eggs, *C. capitata, to* the parasitoid were evaluated. One of these was the use of guavas with natural infestation, a method used until the fifth generation, as described above, and guavas with perforations (5 mm in diameter X 3 mm deep) in the skin, where *C. capitata* eggs were placed.

The treatments consisted of a transparent plastic cage (3.8L) with nylon mesh on the top and small perforations on the sides to increase ventilation, as well as food (bee honey) and water ad libitum, containing ten pairs of *F. arisanus* between 10 and 20 days old. The guavas were exposed to parasitism for 24 hours. They were then transferred to plastic jars (500 mL) sealed with fleece, containing a 2 cm layer of vermiculite at the bottom, which served as a substrate for pupae.

These pots were kept for twelve days in an air-conditioned room (24±2°C, 50±10% RH and 14 h photophase). After this period, the pupae were transferred to plastic pots (250 mL) until the adults emerged. Five replications were carried out for seven consecutive days.

The parameters evaluated were: number of pupae, parasitism rate [no. parasitoids/(no. parasitoids + no. flies)]*100 and parasitoid sex ratio [no. females/(no. females + no. males)].

3.2.1 Parasitism of *F. arisanus* on eggs of *C. capitata*, bisexual strain, on different types of oviposition substrates.

Five types of oviposition substrates (treatments) were tested: T1 - water-agar block (WAB) plus 4 drops of yellow annelin; T2 - WAB plus 4 drops of yellow annelin and guava extract in a ratio of 3 parts solution to 1 part extract; T3 - WAB with guava extract in a ratio of 3:1; T4 - guavas in the initial stage of ripeness, cut lengthways and ten perforations were made in each half (5 mm diameter X 3 mm depth), and T 5 - BAA.

The treatments with agar-water blocks used a solution of 0.52 g of agar per 100 mL of water, which was used as a standard for preparing all the agar treatments. In treatments 1, 2, 3 and 5, the solution was poured into Petri dishes (6.0 x 1.5 cm) forming a block of agar-water, on which a tissue was placed and 100 *C. capitata* eggs were deposited (Figure 1). In treatment 4 (guava with perforations in the peel), 10 *C. capitata* eggs were deposited in each hole, totaling 100 eggs/fruit. The agar substrates were called oviposition units (OU).

Figure 1 - *Ceratitis capitata* eggs deposited on the agar-water block in a Petri dish.

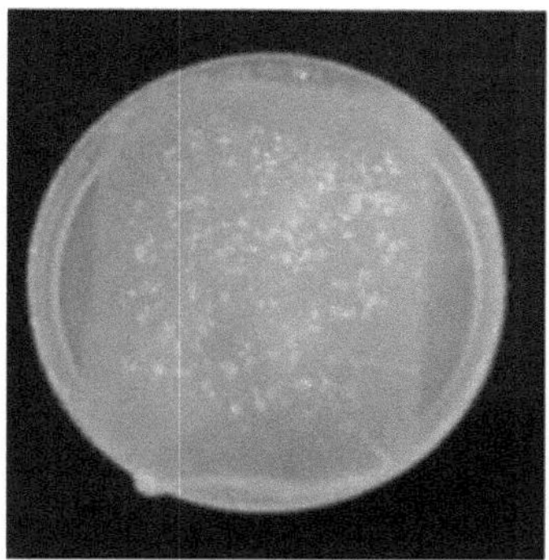
Source: Author, 2013

There were 10 replicates for each treatment. Each replication consisted of a transparent plastic "baleiro" cage (3.8L), with nylon mesh on the top and small perforations on the side to increase ventilation, as well as food (bee honey) and water ad libitum, containing five pairs of *F. arisanus* aged between 10 and 20 days. The oviposition units (OP) and the perforated guavas, each containing 100 eggs, were exposed to parasitism for 24 hours.

After parasitization, the eggs were removed from the guava and the agar blocks using a pick and inoculated onto 200g of artificial diet for *C. capitata* larvae (DAMASCENO, 2013) in Petri dishes (13 cm in diameter) for larval development. The plates were individualized in plastic jars (500 mL), containing a 2 cm layer of vermiculite at the bottom, which served as a substrate for pupation, and sealed with fleece.

These pots were kept for twelve days in an air-conditioned room (24 ± 2°C, RH 50 ± 10% and 14 h photophase) until the pupae formed. The pupae were then transferred to plastic pots (250 mL) until the adults emerged.

The parameters evaluated were: number and weight of pupae (mg), parasitism rate [no. parasitoids/(no. parasitoids + no. flies)]*100, emergence rate of flies (no. flies/no. pupae) *100 and parasitoids (no. parasitoids/no. pupae)*100, parasitoid sex ratio [no. females/(no. females + no. males)] and length of the parasitoid development cycle (egg-adult).

3.2.2 Parasitism by *F. arisanus* on two strains of *C. capitata*, bisexual and mutant *tsl* Vienna 8.

To synchronize the reproductive cycle of the bisexual and *tsl* Vienna 8 mutant strains of *C. capitata* and start the experiment, a cage was set up for each strain of *C. capitata*, following the same methodology described in section 3.2 for rearing the host.

The experiment was carried out using *C. capitata* eggs of the bisexual and *tsl* strains (treatments), with six replicates each. The replication consisted of a transparent plastic "baleiro" cage (3.8L), with nylon mesh on the top and small perforations on the side to increase ventilation, as well as food (bee honey) and water "ad libitum", containing five pairs of *F. arisanus* between 10 and 20 days old.

Guavas at an early stage of ripeness were cut lengthways and ten holes were drilled in each half (5 mm diameter X 3 mm depth), where ten *C. capitata* eggs were deposited per hole, totaling 100 eggs/fruit (100 eggs/repetition) for each strain.

The OUs were exposed to parasitism for 24 hours. The pieces containing the eggs were then carefully cut off and deposited on 200 g of larval diet for *C. capitata* (DAMASCENO, 2013) in Petri dishes (15 cm 0). The pupae were placed in plastic jars (1000 mL) containing a 2 cm layer of vermiculite and kept in air-conditioned rooms (24±2°C, RH 50±10% and 14 h photophase) until the pupae formed. The pupae were then transferred to plastic pots (250 mL) until the adults emerged.

The parameters evaluated were: number and weight of pupae (mg), parasitism rate [no. parasitoids/(no. parasitoids + no. flies)]*100, fly emergence rate (no. flies/no. pupae) *100 and parasitoid emergence rate (no. parasitoids/no. pupae)*100 and parasitoid sex ratio [no. females/(no. females + no. males)].

Experimental design and statistical analysis

The design of the strain experiments was a randomized block with 6 replications, carried out over 5 consecutive days, while the oviposition substrate experiment was carried out in 5 randomized blocks with 2 replications. The data was submitted to analysis of variance (ANOVA) and the means were compared using the Tukey test at 5% probability. The mean and standard error were determined for each variable. The statistical program SPSS version 14.0 Evaluation was used.

3.3 Results and Discussion

Adaptation of the *F. arisanus* colony in Brazil.

The pupal viability of the generations analyzed averaged 85.43 ± 6.1% (Table 1), which shows the good management of the colonies of the host, *C. capitata,* and the parasitoid, *F. arisanus.*

The data from the rearing of *F. arisanus* (Table 1) showed that the parasitoid population had adapted to laboratory conditions. The established colony showed an average parasitism rate of 28.83 ± 7.3% in the first five generations, higher than the average obtained in the rearing of this parasitoid on *B. dorsalis* in Hawaii, which was between 14 and 20.8% (BAUTISTA et. al, 1999).

The sex ratio was variable in the five generations evaluated, with an average of 0.45 ± 0.06. According to Navarro (1998), one of the parameters indicating a good mass rearing method is the recovery of a greater number of female parasitoids in the progenies. Although the average was less than 0.5, it is important to note that the species is adapting to the new host, as it was recently imported from Hawaii, USDA/ARS, where it was reared on *B. dorsalis* eggs.

Table 1. Breeding of *Fopius arisanus* on *Ceratitis capitata* eggs in guavas with natural infestation, from April to September 2013.

Generations	Pupae recovered (n°)	Pupal viability of *C. capitata* and *arisanus* (%)	Emerged *F.* flies (n°)	Emergency *C. capitata* (%) ■	Emerged *F. arisanus* adults (n°)		Parasitism rate (%)	Sex ratio of *F. arisanus*
					Males	Females		
1	664	100	311	46,84	138	215	53,16	0,61
2	2.843	86,98	1.514	53,25	408	551	38,78	0,57
3	2.837	82,93	1.387	71,02	556	410	17,61	0,42
4	4.860	63,74	2.614	53,78	338	146	15,62	0,30
5	2.016	93,50	1.527	75,74	231	127	18,99	0,35
Average	2644±664,9	85,43±6,1	1470±364,9	60,12±5,5	334±72,0	289±82,2	28,83±7,3	0,45±0,06

The development period of *F. arisanus* (egg-adult) was on average 25.8 ± 0.5 days for males (n=1671) and 26.4 ± 0.4 days for females (n=1449), in the five generations analyzed. Zenil et al. (2004) reported the emergence of this parasitoid on the same host at average intervals of 23.0 ± 0.5 days for males (n=1671) and 26.4 ± 0.5 days for females (n=1449) in the five generations analyzed.

0.7 days for males and 24.0 ± 0.7 days for females. In *Anastrepha* spp., the same author describes a significantly longer average development time of 27.4 ± 0.5 days in *A. ludens* and 27.5 ± 0.4 days in *A. serpentina*. It is important to note that the parasitoid's development period is a direct way of assessing the quality of the host (ROITBERG et al., 2001). The shortest development period is considered the best growth path, as better quality hosts allow parasitoids to develop in shorter periods of time (SEQUEIRA; MACKAUER, 1993; ROITBERG et al., 2001).

Breeding of *F. arisanus* in guavas with natural infestation compared to artificial infestation of *C. capitata* eggs

A total of 117 ± 29.2 pupae were recovered from naturally infested guavas and 35 ± 20.0 from artificially infested guavas. Based on these values, it was found that the number of eggs available for parasitism in naturally infested guavas is much lower than in artificially infested guavas.

The viability of the eggs offered in the artificial infestation was 83.8% ± 1.49, which indicates the good quality of the host eggs offered. Despite this, parasitism, at 11.54% ± 2.4, was lower in this type of infestation compared to 24.65% ± 6.9 in the natural infestation. The sex ratio of the parasitoid's offspring in the natural infestation was 0.76 ± 0.1, higher than in the artificial infestation, which was 0.50 ± 0.1.

According to Vargas et al. (1991), the physical appearance of the host fruit is an important factor for *F. arisanus* females to locate the host habitat, and the species is attracted to light colors such as yellow and white. It is also known that the aromas emitted by ripening fruit and the presence of females of the host fly species also act as important clues used by parasitoids in locating and choosing a host (PURCELL et al., 1994; SEGURA et al., 2012; STHUL et al., 2012; BENELLI et al., 2013). Guava is an extremely attractive fruit for various parasitoids (SILVA, BENTO, ZUCCHI, 2007; STHUL et al., 2012). The lower number of eggs on the fruit in the treatment with natural infestation may be related to the daily fecundity of *C. capitata*, which ranges from 6-12 eggs/female/day. This reproductive parameter is influenced by the larval development substrate and the food consumed during the adult stage (PAPADOPOULUS et al., 2002; JOACHIM- BRAVO et al., 2010; ZANARDI et al., 2011).

The lower rate of parasitism in artificially infested guavas may be related to the difficulty experienced by the females of *F. arisanus in* locating the host eggs placed in the perforations (5 mm in diameter X 3 mm deep) made in the skin of the fruit, since in nature the eggs are located under the skin of the host fruit. According to Wang and Messing (2003), the females of this parasitoid generally insert their ovipositor deep into the host's egg-laying cavity. In this way, eggs laid artificially on the surface of the fruit can be more difficult to detect. Furthermore, natural oviposition can leave oviposition pheromone traces, remains of the host's feces, or induce the release of volatiles in the infested fruit, and all these olfactory clues can be used by the parasitoid to locate the host (ROUSSE et al., 2007; PEREZ et al., 2013).

A greater number of females than males in the progeny is highly desirable in mass breeding, since the females are the "active ingredients" in the biological control applied, as they are the ones who will attack the target pest (Paranhos et al, 2008). The female's choice of suitable host at the time of laying is fundamental to the survival and success of her offspring (SINGER, 1986; RENWICK, 1989). The eggs laid in the perforations (5 mm in diameter X 3 mm deep) in the guava skin were more subject to damage to their structure as a result of human handling. In addition, it could contribute to a greater degree of dryness, compromising the quality of the host. Therefore, the higher sex ratio in guavas exposed to natural infestation suggests selection by the female *F. arisanus to* lay a greater number of eggs that will produce females, since the host eggs would be of better quality, as they would be protected from physical damage and drying out.

3.4.1 Parasitism of *F. arisanus* on eggs of *C. capitata,* bisexual strain, on different types of oviposition substrates.

In the treatments with eggs on agar-water blocks, there were no differences in the parameters total number of pupae, egg-pupa recovery and host emergence rate (Table 3). In treatment 4, in which guava was used as an oviposition substrate, there was a decrease in the total number of pupae obtained, in egg-pupa recovery and in the emergence rate of *C. capitata,* when compared to the other treatments (Table 3).

Table 2 - Total pupae, egg-pupa yield and emergence rate [mean ± SD] of adults, bisexual strain, from eggs exposed to parasitism by *F. arisanus,* in different oviposition units.

Treatments[1]	Parameters		
	Total Pupae (n°)	Egg-pupa yield	Emergency rate
T1	74,60 ± 3,07 a	0,75 ± 0,03 a	98,05 ± 0,28 a
T2	66,80 ± 5,19 a	0,67 ± 0,05 a	97,70 ± 0,60 a
T3	67,40 ± 4,43 a	0,67 ± 0,04 a	96,93 ± 0,71 a
T4	47,70 ± 5,38 b	0,48 ± 0,05 b	84,15 ± 4,66 b
T5	72,80 ± 4,27 a	0,73 ± 0,04 a	97,12 ± 0,61 a

T1 = Agar-water and annelin block; T2 = Agar-water, annelin and guava extract solution; T3 = Agar-water and guava extract solution; T4 = Guava; T5 = Agar-water solution. Averages followed by the same letter in the same column do not differ (Tukey, P>0.05).

F. arisanus was only recovered in treatment 4, with averages of 12.57 ± 4.85 and 12.89 ± 4.93 for the percentages of emergence and parasitism, respectively. The weight of the adult parasitoids did not differ between males and females, with averages of 0.002g ± 0.0004 (n= 5) and 0.003g ± 0.0001 (n= 3), respectively. There was no parasitism by *F. arisanus* in the other treatments.

It can be seen that the type of substrate used had a direct influence on the parasitism process. The negative results obtained in the treatments with the agar-water substrate suggest that the females need to follow clues from the volatiles present in the guava to locate their host and carry out oviposition, since the chemical stimuli emitted by ripening fruit are highly attractive to parasitoids (LIQUIDO, 1991; PURCELL et al., 1994).

The good recovery of pupae, as well as the high emergence of *C. capitata* from the treatments with agar-water blocks, indicate that the procedure used did not compromise the quality of the host in these treatments. Future studies should be carried out in order to adjust a methodology that forces the females to parasitize the eggs without the influence of visual or chemical clues present in the fruit, in order to make mass production of this parasitoid feasible by offering eggs on water blocks as an artificial oviposition substrate, as is already done in Hawaii (MANOUKINS et al., 2010).

3.4.1 Parasitism of *F. arisanus* on two strains of *C. capitata*: bisexual and mutant *tsl* Vienna 8.

The hatching rate of *C. capitata* larvae differed between the two strains, with an average of 82.0 ± 3.0 in the bisexual strain and 56.0 ± 5.64 in the *tsl strain.* Egg-to-pupa recovery was also higher in the bisexual strain with an average of 0.47 ± 0.04, differing from the value obtained in the *tsl strain,* whose average was 0.28 ± 0.02 (Table 3).

The percentage of emergence of the fly's adults did not differ between the strains, with values of 84.80 ± 2.76 and 78.19 ± 2.17 for bisexual and *tsl,* respectively.

With regard to the rate of parasitism and emergence of *F. arisanus*, the highest rates were recorded for the bisexual strain of *C. capitata* (Table 3). There was no difference in the sex ratio of the parasitoid's offspring between the two strains studied (Table 3).

Table 3 - Parasitism rate, emergence and sex ratio [mean ± standard error] of *Fopius arisanus* reared on *Ceratitis capitata*, bisexual strain and *tsl Vienna 8* mutant. Temperature 24 ± 2°C, 50 ±10% RH and 14 hours of photophase.

Parameters	Host		-F	P
	C. capitata bisexual	C. capitata tsl-Vienna 8		
Egg-pupa yield	0,47 ± 0,04 a	0,28 ± 002b	4,23	0,00
Parasitism rate	11, 45 ± 2,92a	3,19 ± 1,13 b	6,32	0,02
Emergency rate	11,11 ± 2,89 a	2,03 ± 0,78 b	9,35	0,00
Sexual Reason	0,67 ± 0,06 a	0,67 ± 0,15 a	0,82	0,38

Different letters on the same line indicate a significant difference between the values using the Student's t-test.

The weight of the adult parasitoids was similar in the two strains and did not differ between males and females of each strain, with averages of 0.005g ± 0.0004 and 0.004g ± 0.0006 for bisexual and *tsl,* respectively.

According to Pratissoli and Oliveira (1999), variations in parasitism may be related to the type of host (preferential or preferential).

alternative), origin of the natural enemy strain, generation in which the colony is kept in the laboratory, environmental conditions of temperature, relative humidity and photoperiod, among others, used for the multiplication of natural enemies.

Determining the most suitable host is essential, as a host may be nutritionally inadequate or insufficient for the complete development of the parasitoid, influencing its sex ratio, size, development time, fecundity and longevity (VINSON, IWANTSCH 1980; VAN ALPHEN, JERVIS 1996). Generally, higher quality hosts allow the development of larger and more competitive parasitoids (CHOW, MACKAUER, 2001).

The best results regarding the multiplication of the parasitoid on the bisexual strain of *C. capitata* indicate that this host is more suitable for rearing *F. arisanus*. The adaptation of the parasitoid to the alternative host is an important stage in achieving success in biological control programs. This egg parasitoid is normally reared on *B. dorsalis* at the USDA-PBARC in Hilo, Hawaii. However, before this specimen was sent to Brazil in November 2012, it was reared for a few generations on *C. capitata* eggs, since Brazil is free of the fruit fly species *B. dorsalis.*

In Brazil, the best fly species to use as a host is *C. capitata,* since the *Anastrepha* species present in Brazil are generally not good hosts for *F. arisanus* (CANSINO et al., 2008; MONTOYA et al, 2009) and the carambola fly, *Bactrocera carambolae* (Diptera: Tephritidae), the only species of this genus in Brazil, is restricted to the state of Amapà, and its breeding, even for research purposes, is prohibited elsewhere in Brazil. Therefore, further studies are needed to evaluate other factors that may interfere in the successful multiplication of this exotic parasitoid on *C. capitata,* such as: egg age, egg density, parasitoid and exposure time to parasitism.

4 Use of irradiated eggs (X-rays) of *Ceratitis capitata* in the rearing of *Fopius arisanus* (Hymenoptera: Braconidae).

4.1 Introduction

Ceratitis capitata Wied. 1824 (Diptera: Tephritidae) is the only species of this genus of fruit flies that occurs in Brazil and its presence was first noted in 1901 (IHERING, 1901). Also known as the Mediterranean fruit fly or moscamed, it is native to equatorial Africa and is distributed on all five continents. It is considered one of the most harmful fruit fly species in the world, as it has great ecological and evolutionary plasticity, adapting quickly to new hosts and environments (MALAVASI, 2009). It is widely distributed in South America and Brazil, occurring from the state of Ceará to Rio Grande do Sul. For these reasons, the Mediterranean fruit fly is of great quarantine importance in the international fresh fruit trade.

C. capitata eradication programs have already been carried out or are underway in several countries. In the countries that live with the pest, the export of *C. capitata* host fruit to the United States and Japan has only occurred through the formation of free zones or the application of quarantine treatments (RAGA et *al.*, 1996).

Widely used until 1984, post-harvest quarantine treatments based on fumigants (ethylene dibromide, methyl bromide and phosphine) are being banned worldwide due to their undesirable characteristics, such as toxicity to humans and plants (RAGA et *al.*, 1996). For this reason, control through physical treatments such as food irradiation has been the most widely researched and used. Examples of this are treatments using heat, cold or ionizing radiation.

The use of the sterile insect technique (SIT) relies on ionizing radiation to sterilize the insects. In the 1950s, when CIT programs for *Cochliomyia hominivorax* (Coquerel) began in the United States, the source of ionizing radiation was X-rays, but only for laboratory research and small releases of sterile insects in the field, as it was very expensive. As a result, the alternative became sterilizing insects with gamma radiation from Co^{60} (LINDQUIST, 1955).

Recently, new equipment produced by RadSource Technologies, GA, USA, has changed this scenario and made it possible to use X-rays as a viable technology for sterilizing insects on an industrial scale. The new X-ray technology in TIE is due to a special ampoule with gold wires that emits a 4 pi radiation field similar to a light bulb. The dose rate can vary from 18 to 70 Gy/min, which makes it possible to sterilize 20 L of pupae every 3 to 5 minutes (WALDER, MASTRANGELO, 2011).

In addition to sterilizing insects used in BIT programs, X-rays can be used in biological control programs to multiply parasitoids by applying specific doses of ionizing radiation at different stages of development of the host, immediately before exposure to parasitism, in such a way as to prevent the emergence of the host insect's adult without, however, compromising the development of the parasitoid inside the host. In addition, irradiation has been shown to induce changes in the host's immune system, weakening the natural defense against the introduction of foreign bodies, thus acting synergistically on the efficiency of parasitism (COSTA, et *al.*, 2008).

In this way, the use of radiation favors the production of parasitoids and does not cause negative effects on their development (SIVINSKI; SMITTLE, 1990; CANCINO et *al.*, 2010).

The use of irradiated *C. capitata* larvae in the multiplication of the parasitoid *Diachasmimorpha longicaudata* (Hymenoptera: Braconidae) is aimed at "clean" Massai breeding, meeting the objectives of applied biological control programs, where parasitoids are periodically released into the field, without the risk of releasing the flies that are pests of fruit trees.

Similarly, the exotic egg parasitoid *Fopius arisanus* (Hymenoptera: Braconidae), imported in November 2012 by Embrapa Meio Ambiente, could theoretically be multiplied on irradiated *C. capitata eggs*. To this end, the aim of this work was to test different doses of X-rays applied to the eggs of two strains of *C. capitata* (Bisexual and mutant *tsl* Vienna 8), in order to define a dose of radiation that prevents the emergence of adults of this species of fruit fly, but at the same time does not compromise the development of the immature stages of the parasitoid inside.

4.2 Material and Methods

The insects used came from the rearing of two strains of *C capitata*: bisexual and mutant *tsl* Vienna 8 (*temperature sensitive lethal*) and the experiments were conducted under controlled environmental conditions (temperature of 25 ± 2 °C, relative humidity (RH) of 50 ± 10%, and photophase of 14 h).

Breeding *C. capitata* (bisexual and *tsl* Vienna 8 mutant)

An adult cage (25 x 40 x 35 cm) with an aluminum frame, acrylic base and voal fabric sides was set up for each of the strains. 100 mL of pupae were placed in each cage, corresponding to approximately 4,200 adults (1   ♀:1 ♂), a container containing water and another containing an adult diet (3:1, sugar: hydrolyzed protein - Biones®) (SILVA

NETO et *al.*, 2012). To oviposit, the female flies insert the ovipositor into the lateral fly tissue and oviposit. The eggs fall by gravity into trays of water placed below the cages, from where they are collected every 24 hours and placed in plastic trays on an artificial diet for larvae developed by DAMASCENO (2013).

Rearing the parasitoid *F. arisanus* on the bisexual strain of *C. capitata*

Adults of *F. arisanus* were placed in cages (30 x 30 x 30 cm) with an aluminum frame, fixed acrylic bottom, top and sides with fiberglass mesh (1 mm mesh), containing water and bee honey "ad libitum".

For multiplication, guava fruits (*Psidium guajava L.) were* used, at an early stage of ripeness, with perforations in the skin 5 mm in diameter and 2 mm deep, containing 24 h eggs of *C. capitata*. These parasitism units were then offered to *F. arisanus* adults, from the 9th day of age, for 24 hours.

After parasitization, the eggs were carefully removed by cutting off the pieces of guava and inoculating them on an artificial diet for *C. capitata* larvae (DAMASCENO, 2013), placed in Petri dishes (13 cm in diameter) for larval development. The plates were individualized in plastic jars (500 mL) sealed with fleece, containing a 2 cm layer of vermiculite at the bottom, which served as a substrate for pupation.

These pots were kept for 15 days in an air-conditioned room (24 ± 2°C, RH 50 ± 10% and 14 h photophase) until the pupae formed. The pupae were then transferred to plastic pots (250 mL) until the adults emerged.

4.2.1 Doses of X-ray applied to the eggs of *C. capitata*, bisexual and *tsl* Vienna 8 mutant strains, to prevent the emergence of adults.

C. capitata eggs approximately 24 hours old were subjected to the following doses of X-rays (treatments): control (0 Gy), 5, 15, 25, 35 and 45 Gy. Before irradiation, the eggs were quantified on moistened black cloth in Petri dishes (150x15mm) and placed on moistened filter paper slides (4x4cm) (Figure 3A). Afterwards, the slides with the eggs were rolled up, placed inside a microtube (2 ml) (Figure 3B) and sent to the X-ray machine.

Figure 2 - Petri dish containing four units of filter sponge fabric with 2000 *Ceratitis capitata* eggs, 500 in each (A). The eggs are placed on moistened filter paper inside the microtube (B).

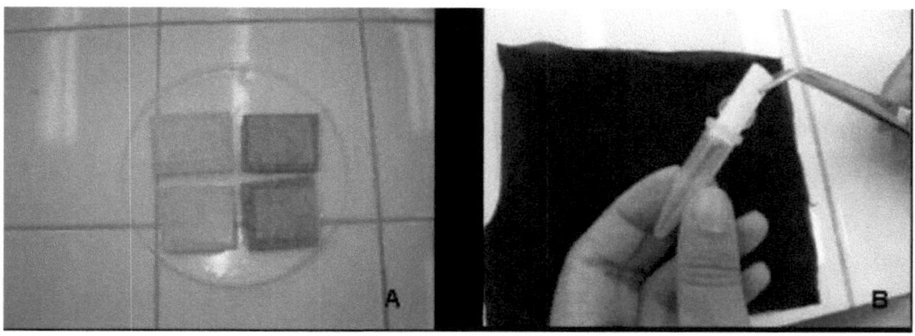

Source: Author, 2013.

Four replicates were used for each treatment (=dose), each with 500 eggs, totaling 2000 eggs/treatment. To check the hatching rate of the eggs that were going to be subjected to radiation, a quality control was carried out in Petri dishes (90x15mm), consisting of five replicates of 100 eggs each, for each strain tested (Bisexual *and tsl* Vienna 8 mutant).

The biological material was irradiated in an X-ray source, model Rad Source RS 2400 (Figure 3), equipment produced in the USA for use in insect sterilization. The machine has five cylinders of 4 L each, which rotate around the radiation field. The dose rate used was approximately 2.754 Gy/min. The irradiator was operated with a voltage of 120 Kv and a current of 18 mA. Two microtubes containing eggs, one from each strain, were placed equidistantly in the center of a cylinder for each dose tested.

Figure 3 - Rad Source X-ray machine (RS-2400).

Source: Author, 2013.

After the irradiation process, the microtubes containing the eggs were taken back to the laboratory and the eggs were inoculated onto an artificial diet (DAMASCENO, 2013), placed in Petri dishes (15 cm 0).

The four replicates (250 g plates with larval diet + 500 eggs/replicate) of each treatment (dose) were placed in plastic trays (57 x 37 x 2.7 cm) and covered with fleece to prevent

31

drosophila infestation.

The trays were kept in an air-conditioned room (T = 24 ± 2°C and RH = 50 ± 10%) (Figure 4A). Six days after inoculation, when the larvae inside the diet were already in the third and last larval instar, each Petri dish was transferred individually to a tray (38 x 27 x 10 cm) containing a layer of vermiculite on the bottom as a pupae-forming substrate (Figure 4B). After eight days, the pupae recovered from each treatment and repetition were quantified, volumetrized (ml) and the egg-pupa yield was calculated (number of pupae recovered / number of eggs inoculated). Next, pupal weight (mg), emergence (%E) (number of flies that emerged / number of pupae)*100 and flies (%V) (number of flies that flew / number of pupae)*100 were evaluated. The weight of the samples was measured on a precision analytical balance, model AR3130, with a deviation of 0.001g.

Figure 4 - Tray containing Petri dishes with eggs inoculated on artificial diet (A) and Petri dish placed in a tray containing vermiculite (B).

Source: Author, 2013.

The experimental design used was a factorial design with six treatments and four replications. The data was statistically analyzed using SAS version 9.0. Tukey's test was used to compare the means at a 5% significance level.

4.2.2 Parasitism of *F. arisanus* on irradiated and non-irradiated *C. capitata* eggs

C. capitata eggs from colonies kept at the Moscamed Brasil Biofactory were used.

Every day, for a period of five consecutive days, 1000 *C. capitata* eggs were counted and separated, 500 of which were irradiated and 500 of which were not irradiated.

All batches of 500 eggs, irradiated and non-irradiated, were transferred to moistened filter paper, rolled up and placed individually inside a 50 mL conical tube (Figure 5A). The tubes with eggs to be irradiated were placed in the center of the rice-filled cylinder to even out the radiation received (Figure 5B).

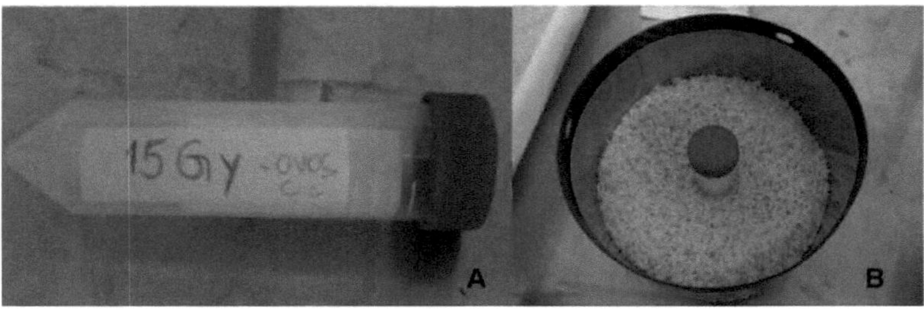

Source: Author, 2013.

The irradiation dose applied was 15 Gy, defined previously (item 4.2.1), which prevented the emergence of *C. capitata* adults.

After irradiation, the tubes containing the eggs were taken back to the laboratory and the eggs were carefully distributed on black cloth over foam moistened with water in Petri dishes for counting.

A guava fruit (*Psidium guajava* L) at an early stage of ripeness was used as the oviposition unit (OU*)*. The fruit was washed with running water and neutral detergent, then placed in a solution containing one tablespoon of sanitary water per 1L of water and left there for about 5 minutes. They were then rinsed with plenty of running water and dried with paper towels. Once this stage had been completed, the guavas were cut lengthways, taking advantage of only half the fruit, in which five holes were made in the skin, measuring 5 mm in diameter by 2 mm deep, distributed equidistantly on the surface of the fruit. Ten *C. capitata* eggs were deposited in each hole, totaling 50 eggs/UO (50 eggs/repetition).

The eggs prepared according to the methods described above were transferred to the cages containing *F. arisanus*. Six cages (replicates) were set up for each treatment (irradiated eggs and non-irradiated eggs), each containing five pairs of the parasitoid, aged between 10 and 20 days. The cages, adapted from 3.6 L plastic tubs, had nylon mesh on the top and small perforations on the sides to increase ventilation (Figure 7), with water and honey "*ad libitum*".

Figure 6 - Baleiro cage containing five pairs of *Fopius arisanus,* one UO with 50 eggs, food and water.

Source: Author, 2013.

The OUs with the eggs were offered for a period of 24 hours. After exposure to parasitism, the eggs were removed from the OU by cutting around the egg holes with a knife and transferring them to 250g of artificial diet for *C. capitata* larvae, in Petri dishes 13 cm in diameter, for the larval development of *C. capitata*, individually for each treatment and repetition.

The plates were individualized in 500 mL jars sealed with voal cloth (Figure 7), which were then covered with a two-centimetre layer of vermiculite to form the pupae. The whole experiment was carried out in an air-conditioned room, with temperatures of 25 ± 2° C and an air RH of 60 ± 10%.

Figure 7 - Pots containing 250 g of artificial diet for *Ceratitis capitata* larvae with small pieces of guava containing eggs (irradiated or not), shortly after exposure to parasitism by *Fopius arisanus*.

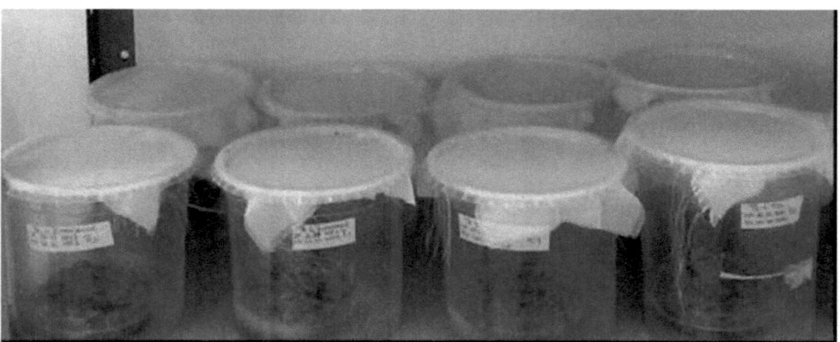

Source: Author, 2013.

The parameters evaluated were: number and weight of pupae (mg), emergence rate of flies and parasitoids (no. flies/no. pupae) *100 and (no. parasitoids/no. pupae)*100, parasitoid sex ratio [no. females/(no. females + no. males)] and length of the parasitoid development cycle (egg-adult).

The experimental design was randomized blocks, with two treatments: irradiated and non-irradiated eggs. Six replications were carried out for five consecutive days. The results were analyzed using the statistical program SPSS version 14.0 Evaluation. The data was subjected to analysis of variance and when there were differences between treatments, the analysis was carried out using mean comparison tests, using the Student's t-test to determine statistical differences. The means and standard error were determined for each treatment.

4.3 Results and Discussion

Doses of X-ray applied to the eggs of *C. capitata*, bisexual and *tsl* Vienna 8 mutant strains, to prevent the emergence of adults

The studies showed that there was a difference between the treatments for egg-pupa yield (F=20.89; P<0.001 and F=25.03; P<0.001), pupal weight (F=39.98;P<0.001   and F=101.48; P<0.001) and adult emergence (F=6.590.51; P<0.001 and F=367.71; P<0.001), in the two strains tested, bisexual and mutant *tsl* Vienna 8, respectively.

The average egg-pupa yield for the bisexual strain obtained at a dose of 5 Gy was similar to the control (non-irradiated) and at the other doses the averages were lower and different from the control (Fig. 8). In the *tsl strain*, the egg-pupa yield was similar to the control up to a dose of 25 Gy, while at higher doses the yields were lower and different (Fig. 8). The egg stage in insects is generally more sensitive to the effects of irradiation, as these are inversely proportional to the degree of cell differentiation (CANCINO et al. 2009a), so it is assumed that in treatments using higher doses of X-rays, the low egg-pupa yield is due to the greater vulnerability of 24-hour eggs.

It was observed that from 35 Gy onwards, radiation had a negative influence on the development of *tsl* larvae, reducing the number of pupae formed. From 35 Gy onwards, the doses of irradiation proved to be totally lethal to the larvae that would give rise to females (Table 4), as there were no white pupae (white pupal mutation of the *tsl* Vienna 8 strain) from this dose onwards. This suggests a greater sensitivity of the embryos of females of the *tsl* strain, which was to be expected as the mutation is located on the X chromosome (CACERES, 2002).

Figura 8   - Egg-pupa yield of bisexual and *tsl* Vienna 8 mutant strains of *Ceratitis capitata* subjected to different doses of X-rays (control = 0, 5, 15, 25, 35 and 45 Gy).

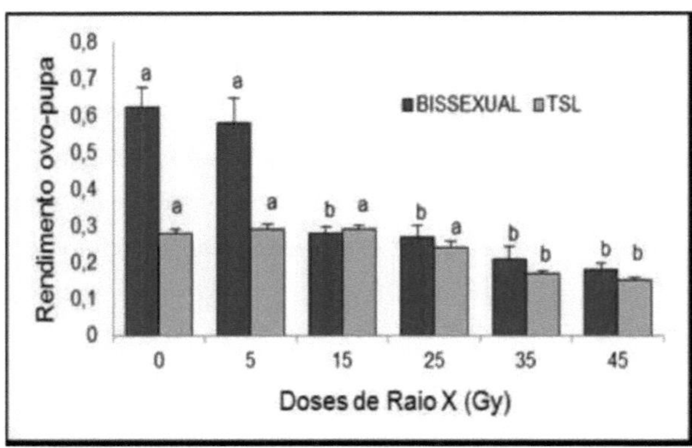

Source: Author, 2013.

Table 4 - Average percentage recovery of white and brown pupae (± EP) from eggs of *Ceratitis capitata* strain *tsl* Vienna 8, subjected to doses of 0 (non-irradiated = control), 5, 15, 25, 35 and 45 Gy of X-rays.

Dose (Gy)	Brown pupae* (%)	White Pupae** (%)
0 (control)	47,91 ± 2,51	52,09 ± 2,51
5	50,17 ± 5,11	49,83 ± 5,11
15	75,21 ± 1,29	24,79 ± 1,29
25	100,00 ± 0,0	0 ± 0,0
35	100,00 ± 0,0	0 ± 0,0
45	100,00 ± 0,0	0 ± 0,0

* Males originate from the *tsl* Vienna lineage 8

** Females originate in the *tsl* Vienna lineage 8

The pupal weight averages for the bisexual strain showed that the 5 Gy dose was similar to the control (non-irradiated) and for the other doses the averages were lower than the control (Fig. 9). In the *tsl,* the weight of the pupae decreased as the radiation dose increased, starting at 15 Gy (Fig. 9). This result corroborates that observed by Cancino et al. (2009a) in their study, which showed an inverse relationship between the dose of irradiation applied and the weight of the pupae obtained.

Figura 9   - Average weight of pupae of the bisexual and *tsl* Vienna 8 mutant strains of *Ceratitis capitata* from eggs subjected to doses of 0 (non-irradiated = control), 5, 15, 25, 35 and 45 Gy of X-rays.

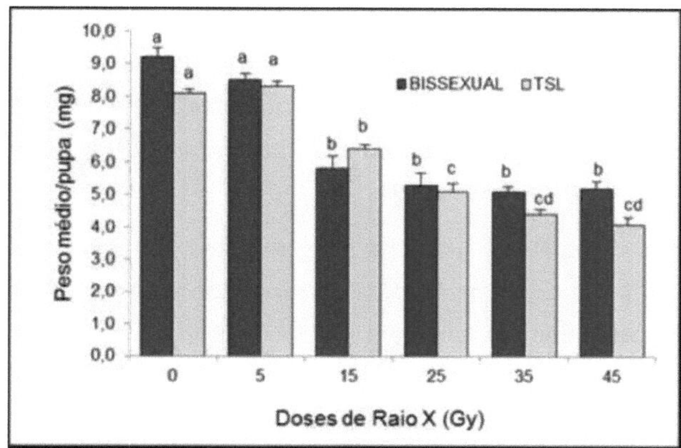

Source: Author, 2013.

It is therefore necessary to carry out studies analyzing different ages of host eggs.

Pupae lose weight over time as a result of the energy expended during the metabolic process. According to FAO/IAEA (1999), a standard weight for the *C. capitata* species is 8.0 mg (with an acceptable minimum of 7.5 mg) and 7.2 mg (acceptable minimum of 6.7 mg) for the bisexual and *tsl* Vienna 8 mutant strains, respectively. Therefore, in the bisexual strain, the standard weights were achieved in the control and at a dose of 5 Gy. In the *tsl strain*, however, only the pupae that had their eggs subjected to doses of 35 and 45 Gy did not reach the minimum acceptable weights.

The average emergence of *C. capitata* adults for the two strains, bisexual and *tsl*, at a dose of 5 Gy was lower and different from that observed in the control (non-irradiated) and at the other doses emergence was zero or close to it (Fig. 10).

Figura 10 - Average emergence of adults of the bisexual and *tsl* Vienna 8 mutant strains of *Ceratitis capitata* from eggs subjected to doses of 0 (non-irradiated = control), 5, 15, 25, 35 and 45 Gy of X-rays.

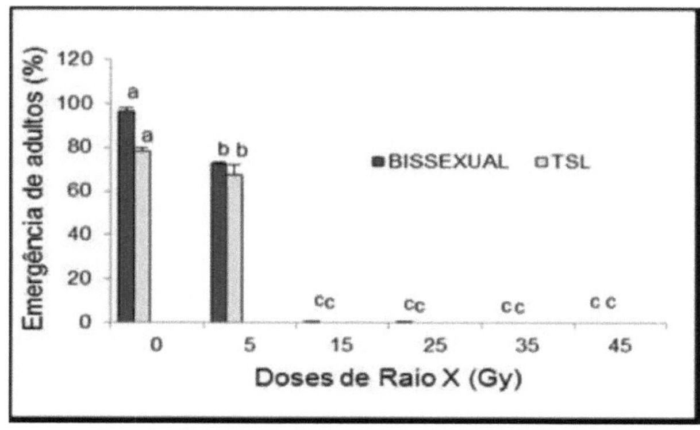

Cancino et al. (2009b) applied different doses of gamma rays to *Anastrepha spp.* and *C. capitata* larvae. In all fly species, emergence decreased as the irradiation dose increased (between 5 and 20 Gy). Total suppression of emergence in *A. obliqua* and *C. capitata* occurred at doses of 30 Gy.

Among the doses of X-rays tested in the Rad Source machine, operated at a voltage of 120 Kv and a current of 18 mA in these experiments, 15 Gy showed the best results for irradiating *C. capitata* eggs, as it prevented the emergence of *C. capitata* adults. At this dose, the egg-pupa yield and pupal weight obtained from the two strains studied, bisexual and the *tsl* Vienna 8 mutant, were similar. In the bisexual, these values decreased compared to the control, but in the *tsl* Vienna 8 mutant, the egg-pupa yield was similar to the control, with only the weight of the pupae formed decreasing.

The pupal weight obtained with 15 Gy was lower than the quality control standard for *C. capitata* (FAO/IAEA/USDA, 2003). However, further studies should be carried out to see if the lower pupal weight could negatively affect the quality of the parasitoid produced on this host.

In addition, further studies should be carried out using intermediate doses between 5 and 15 Gy in order to see if it is possible to obtain a dose that results in a higher egg-pupa yield, with low emergence of hosts (flies) and a good parasitism rate. All these studies are essential for establishing mass rearing protocols for this exotic parasitoid on irradiated *C. capitata* eggs in Brazil.

Parasitism by *F. arisanus* on irradiated and non-irradiated eggs of *C. capitata*, bisexual strain.

There was no difference between the two treatments in the average hatching rate of *C. capitata* larvae, with 85.5 ± 2.10% for irradiated eggs and 92.8 ± 0.85% for non-irradiated eggs. These values indicate that the eggs used in the experiment were of good quality and that the low recovery of pupae was due to the irradiation of the eggs, which caused mortality in the larval stage.

Table 5 - Average number of pupae, flies and parasitoids emerged and parasitism rate from irradiated (15 Gy) and non-irradiated eggs of *Ceratitis capitata*.

Treatment	Total Pupae	No. of *C. capitata*	Emergence *capitata*	*C.* No. of *arisanus*	*F* Parasitoid emergence (%)	Reason Sexual of *F. arisanus*

| 0 Gy | 2.301 | 2.101 | 92,87 ± 1,76 a | 49 | 1,37± 0,69 a | - |
| 15 Gy | 146 | 6 | 1,71 ± 1,24 b | 2 | 2,72 ± 1,24 a | 0,48 ± 0,12 |

Averages followed by the same letter in the column do not differ according to the Tukey test (0.05).

This result with *C. capitata* eggs is in agreement with the observations made by Sivinski and Smitle (1990) who studied the effects of Gamma radiation on the development of *A. suspensa and* the subsequent development of the parasitoid *D. longicaudata.* These authors found that fruit fly larvae subjected to gamma radiation at doses that prevent the emergence of the fly's adult can be used for parasitism, without affecting the development of the parasitoid.

The emergence rate of parasitoids on irradiated and non-irradiated eggs showed no significant difference between the two treatments (Table 5).

Costa et al. (2008) identified a significant increase in the parasitism of *D. longicaudata* on irradiated larvae *tsl* Vienna 8. Walder and Costa (2003) reported that due to its ionizing action, gamma radiation induces alterations in the physiological process that affect the normal defence mechanisms of *C. capitata* larvae, which contributes to the efficiency of parasitism.

The results obtained by Cancino et al. (2009b) with gamma rays on *Anastrepha* spp. and *C. capitata* larvae show that the use of an irradiated host has no negative effect on the development process of the parasitoid *D. longicaudata.* This indicates the feasibility of using irradiation for mass rearing of fruit fly parasitoids.

There was less recovery of pupae (Table 5) and emergence of *C. capitata* in the treatment with eggs irradiated with 15 Gy (X-rays) compared to the control (non-irradiated eggs). The dose of 15 Gy proved to be effective in preventing the emergence of *C. capitata* (Table 5). There was a low number of emerged adults (flies and parasitoids) in the treatment from irradiated eggs that were subsequently exposed to parasitism.

The average weight of 0.003 ± 0.0004g for males and 0.003 ± 0.0002g for females of *F. arisanus* did not differ in the control (non-irradiated eggs) and could not be compared in the treatment with irradiated eggs (15 Gy), due to the non-emergence of females.

The low emergence and parasitism may be due to the lack of adaptation of this host, *C. capitata,* to *F. arisanus.* However, its adaptation to the new host may occur after several generations and new studies should be repeated in the future.

In the control, in which the eggs exposed to parasitism were not irradiated, there was a high emergence of flies but a low emergence of parasitoids. This indicates that there was no overparasitism, because when this happens, the emergence of both hosts and parasitoids decreases (MONTOYA et al., 2009).

In this case, the low parasitism rate obtained in the two treatments, irradiated and non-irradiated eggs, may be related to a high host: parasitoid ratio (egg: female), which may have been much higher than necessary to obtain a good parasitism rate. In addition, the eggs placed in the perforations (5 mm in diameter X 3 mm deep) made in the guava skin would have been subject to drying out and a greater degree of manipulation, which could have compromised the quality of the host and, consequently, the choice and good development of the parasitoid.

The sex ratio of the progeny from non-irradiated eggs (control) showed an average of 0.48 ± 0.12, a value close to 0.50 which is considered ideal, according to Navarro (1998). In the progeny from irradiated eggs, it was not possible to calculate the sex ratio due to the low number of parasitoids that emerged.

Hepdurgun et al. (2009) pointed out that the application of gamma rays to *C. capitata* larvae had no effect on the sex ratio of the parasitoid *Psyttalia (=Opius) concolor* (Hymenoptera: Braconidae).

Further studies should be carried out, using doses of between 5 and 15 Gy, to see if it is possible to obtain a higher egg-pupa yield, with pupae of good weight, but which prevent the emergence of the hosts (flies). In addition to studies relating to the age of the eggs and exposure time to parasitism.

In industrial insect production (mass rearing), the highest possible productivity must be sought and, to this end, two parameters are extremely important: egg-pupa yield and parasitism rate. In this case, rearing procedures ranging from the collection of eggs from the host fly, age of the host egg to be irradiated and parasitized, radiation dose and exposure time to parasitism must be well established in order to obtain the highest number of pupae with the largest possible progeny of parasitoids.

All these studies are essential in order to establish "clean" Massai breeding protocols for the exotic parasitoid, *F. arisanus*, on irradiated *C. capitata* eggs in Brazil.

5 FINAL CONSIDERATIONS

The multiplication of *F. arisanus* on guavas with artificial infestations of *C. capitata* eggs showed a higher yield compared to natural infestation, because although the parasitism rate was lower, the recovery of parasitoids was higher. Guavas with artificial egg infestation are a semi-artificial breeding method for *F. arisanus.* All artificial substrates using agar-water blocks, with and without colors and odors, were unsuccessful in multiplying this parasitoid. However, new studies with adjustments to the percentage of guava extracts in the blocks, as well as the use of oviposition pheromones from the host fly, *C. capitata,* should be tested.

The dose of 15 Gy of X-rays, in the Rad Source machine and operated at a voltage of 120 Kv and a current of 18 mA, applied to *C. capitata* eggs of the bisexual and *tsl* Vienna 8 mutant strains, allowed the immature stage of the fly to develop into pupae and was enough to almost completely prevent the adults from emerging.

At a dose of 15 Gy, both strains of *C. capitata* were similar with regard to the parameters egg-pupa yield and pupal weight. In this way, both could be used to multiply *F. arisanus.* However, the use of the *tsl* Vienna 8 mutant strain in the multiplication of parasitoids is only justified in biofactories that already use this strain for the production of sterile males, such as the Moscamed Brasil Biofactory, located in Juazeiro-BA, since the cost of energy and labor is higher in the production of this strain than in the bisexual strain.

The parasitism rate in both irradiated and non-irradiated eggs of the bisexual *C. capitata* strain was very low. The irradiation of eggs with 15 Gy (X-rays) of the bisexual strain, followed by exposure to parasitism by *F. arisanus*, for a period of 24 h, had a negative effect on the survival of the larvae, with a large decrease in the egg-pupa yield.

This work is the first stage of studies with *F. arisanus* in Brazil and *C. capitata has* not proved to be an ideal host for multiplication, however, this is the only species available for its rearing, with the exception of the state of Amapà, where it will be possible to multiply on the carambola fly, *B. carambolae.* Future work should focus mainly on intermediate X-ray doses between 5 and 15Gy to be applied to *C. capitata* eggs. In addition, different ages of eggs exposed to irradiation should be studied, as well as the length of time they should be exposed to parasitism, in order to obtain good egg-pupa yields, good pupal weights, low pest emergence and good parasitism rates. These adjustments are essential for the mass rearing of *F. arisanus so* that this exotic egg parasitoid can be successfully used in biological control programs applied to the Mediterranean fruit fly, *C. capitata,* in the Sâo

Francisco Valley and to the carambola fruit fly, *B. carambolae*, in Amapâ.

6 REFERENCES

ALUJA, M. Bionomics and management of *Anastrepha*. **Annual Review of Entomology**, Palo Alto, v. 39, p. 155-178, 1994.

ALUJA, M.; MANGAN, R. L. Fruit Fly (Diptera: Tephritidae) Host Status Determination: Critical Conceptual, Methodological, and Regulatory Considerations. **Annual Review of Entomology**, Palo Alto, v. 53, p. 473-502, 2008.

ALVARENGA, C. D.; SILVA, M. A.; LOPES, G. N.; LOPES, E. N.; BRITO, E. S.; QUERINO, R. B.; MATRANGOLO, C. A. R. Occurrence of Ceratitis capitata Wied. (Diptera: Tephritidae) in papaya fruits in Minas Gerais.

Neotropical Entomology, v. 36, n. 5, 2007.

ALVARENGA, C. D.; GIUSTOLIN, T. A.; QUERINO, R. B. Alternatives for controlling fruit flies. In: VENZON, M.; PAULA JÙNIOR, T J. de;

PALLINI, A. (Coord.) **Alternative technologies for the control of pests and**

diseases. Viçosa, MG: EPAMIG, 2006. chap. 11, p. 227-252.

BAKRI, A.; HEATHER, N.; HENDRICHS, J.; FERRIS, I. Fifty Years of Radiation Biology in Entomology: Lessons Learned from IDIDAS. Annals of the Entomological Society of America, Lanham, v. 98, n. 1, p. 1-12, 2005.

BAUTISTA, R. C.; MOCHIZUKI, N.; SPENCER, J. P.; HARRIS, E. J.; ICHIMURA, D. M. Mass-rearing of the Tephritid fruit fly parasitoid Fopius arisanus (Hyme Noptera: Braconidae). **Biological Control**, v.15, p. 137-144, 1999.

BAUTISTA, R. C., HARRIS, E. J. Effect of fruit substrates on parasitization of tephritid fruit flies (Diptera) by the parasitoid Biosteres arisanus (Hymenoptera: Braconidae). **Environmental Entomology,** v. 25, p. 470-475, 1996.

BENELLI, G., REVADI, S.; CARPITA, A.; GIUNTI, G; RASPI, A; ANFORA, G;

CANALE, A. Behavioral and electrophysiological responses of the parasitic wasp Psyttalia concolor (Szépligeti) (Hymenoptera: Braconidae) to Ceratitis capitata-induced fruit volatiles. **Biological Control**, v. 64, p. 116-124, 2013.

BRAZIL. Federal Superintendence of Agriculture, Livestock and Supply in the State of Pará. Portaria n.94, de 24 de dezembro de 2010, "estabelece as resoluções referentes a recente detecção da praga quarentenaria *Bactrocera* carambolae, Diptera, Tephrifideo (mosca da carambola) na sede do município de Normandia/RR", Diàrio Oficial da

República Federativa do Brasil, Brasilia, 27 dez. 2010. Section 1, p. 47.

BRAZIL. National Secretariat for Agricultural Defense. Ordinance No. 106, of November 14, 1991, "accredits the Quarantine Laboratory of useful organisms for the biological control of pests and others", located at the National Research Center for the Defense of Agriculture CNPDA/EMBRAPA. Official Gazette of the Federative Republic of Brazil, Brasilia, November 20, 1991, Section 1, p. 26246.

BUAINAIN, A. M.; BATALHA, M. O. **Fruit production chain.** Brasilia: IICA/MAPA/SPA, 2007, v.7, 102 p.

CACERES, C.; RAMIREZ, E.; WORNOAYPORN, V.; ISLAM, A.; AHMAD, S. A protocol for storage and long-distance shipment of mediterranean fruit fly (Diptera:

Tephritidae) eggs. I. Effect of temperature, embryo age, and storage time on survival

and quality. **Florida Entomologist**, Gainesville, v. 90, n. 1, p. 103-109, 2007.

CACERES, C. Mass rearing of temperature sensitive genetic sexing strains in the Mediterranean fruit fly (Ceratitis capitata). Genetica, The Hauge, v. 116, p. 107-116, 2002.

CACERES, C; FISHER, K.; RENDON, P. Mass rearing of medfly temperature sensitive lethal genetic sexing strain in Guatemala. In: TAN, K. H. (Ed.). **Areawide management of fruit flies and other major insect pests.** Penang, Malaysia: Universiti Sains Malaysia Press, 2000. p. 543-550.

CAMPOS, W. G.; ARAÙJO, E. R. **Ecology of parasitoid insects and the biological control of pests.** Vertentes , n. 4, p. 79-93, 1994.

CANCINO, J.; RUIZ, L.; LOPEZ, P.; MORENO, F. Mass rearing of parasitoids. *En:* **Moscas de la fruta: Fundamentos y procedimientos para su manejo.** México, D.F. p. 291-306. 2010.

CANCINO, J.; RUiZ, L.; PÉREZ, J.; HARRIS, E. Irradiation of Anastrepha ludens (Diptera: Tephritidae) eggs for the rearing of the fruit fly parasitoids, Fopius arisanus and Diachasmimorpha longicaudata (Hymenoptera:

Braconidae). In: BLOEM, K.; GREANY, P.; HENDRICHS, J. **Biocontrol Science and Technology**, v. 19, n. 1, p. 167-177, 2009a.

CANCINO, J.; RUiZ, L.; LÓPES, P.; SIVINSKI, J. The suitability of Anastrepha spp. and Ceratitis capitata larvae as hosts of Diachasmimorpha longicaudata and Diachasmimorpha tryoni: Effects of host age and radiation dose and implications for quality control in mass rearing. In: BLOEM, K.; GREANY, P.; HENDRICHS, J. **Biocontrol Science and**

Technology, v 19, n. 1, p. 81-94, 2009b.

CARVALHO, R. S. Biocontrol of fruit flies: history, concepts and strategies. **Bahia Agricola**, v. 7, n. 3, p. 14-17, 2006.

CARVALHO, R. S.; NASCIMENTO, A . S. Rearing and use of *Diachasmimorpha longicaudata* for biological control of fruit flies (Tephritidae). In: PARRA, J. R. P.; BOTELHO, P. S. M.; CORRÊA-FERREIRA, B. S.; BENTO, J. M. S. (Eds.). **Biological control in Brazil: parasitoids and predators**. Sâo Paulo: Monole. 2002, p.165-179.

CARVALHO, R. S.; NASCIMENTO, A. S.; MATRANGOLO, W. J. R.; Biological control. In: MALAVASI, A.; ZUCCHI, R. A. (Eds.). **Fruit flies of economic importance in Brazil.** Basic and applied knowledge. Ribeirâo Preto: Holos, 2000. p.113-117.

CHOW A., MACKAUER M. Preference of the aphid parasitoid Monoctomus paulensis (Hymenoptera: Braconidae, Aphidiinae) for different aphid species: female choice and offspring survival. **Biological Control**. v. 20, p. 30-38, 2001.

CHONG, M. Production methods for fruit fly parasites. Proceedings of the Hawaiian **Entomological Society**, v. 18, p. 61-63, 1962.

COSTA, M. de L. Z.; COSTA, K. Z.; ALCARDE, L. D.; SANTANA, Q. B.; KAMIYA, A. C.; BLUMER, L.; PARANHOS, B. A. J.; WALDER, J. M. M. Ionizing radiation as a factor in increasing the parasitism index in the mass rearing of Diachasmimorpha longicaudata (Ashmead) (Hymenoptera: Braconidae) on Ceratitis capitata (Wied., 1824) (Diptera: Tephritidae) strain TSL-Vienna 8. In: SIMPÓSIO DE CONTROLE BIOLÓGICO, 11., 2009, Bento Gonçalves. Technology and environmental conservation: Abstracts. [Bento Gonçalves]: Entomological Society of Brazil: IRGA: Unisinos: Fiocruz, 2009. 1 CD-ROM. **URL: http://ainfo.cnptia.embrapa.br/digital/bitstream/item/17565/1/Beatriiz.pdf**

COSTA, V. A.; BERTI-FILHO, E.; SATO, M. E. Parasitoids and predators in pest control. In: PINTO, A. S.; NAVA, D. E.; ROSSI, M. M.; MALERBO- SOUZA, D. T. (Ed.). **Biological pest control: in practice**. PINTO, A.L. et al. (Org.). Piracicaba: CP 2, 2006. p. 25-34.

CRESONI-PEREIRA, C.; ZUCOLOTO, F. S. Fruit flies (Diptera). In: PANIZZI, A. R.; PARRA, J.R.P. (Ed.). **Insect bioecology and nutrition: basis for integrated pest management**. Brasilia, DF: Embrapa Informaçâo Tecnológica, 2009. p. 733-766.

DAMASCENO, I. C. **Influence of larval diet composition and x-radiation on the quality of ceratitis capitata wiedemann, 1824 (diptera: tephritidae) produced in mass rearing.** 2013. 68 f. Dissertation (Master's Degree in Agricultural Defense) - Federal

University of Recôncavo da Bahia. Cruz das Almas, BA, 2013.

DEBACH, P. 1968. **Control biologico de las plagas de insetos y malas hierbas**. Editora Continental, S.A., Mexico. 927p.

DYCK, V. A.; FLORES, J. R.; VREYSEN, M. J. B.; FERNANDEZ, E. E. R.; TERUYA,

T.; BARNES, B.; RIERA, P. G.; LINDQUIST, D.; LOOSJES, M. Management of Area

Wide Integrated Pest Management Programs that Integrate the Sterile Insect Technique. DYCK, V. A.; HENDRICHS, J.; ROBINSON, A. S. (Ed.). **Sterile Insect**

Technique: Principles and Practice in Area-Wide Integrated Pest Management

2005, XIV, Hardcover ISBN: 1-4020-4050-4, p. 525-542.

DUARTE, A . L.; MALAVASI, A . Quarantine treatments. In: MALAVASI, A.;

ZUCCHI, R. A. (Eds.). **Fruit flies of economic importance in Brazil.** Basic and applied knowledge. Ribeirao Preto: Holos, 2000. p.187192.

EMBRAPA. National Research Center for Environmental Impact Monitoring and Assessment. Referral of processes - protocol for assessing the risk of introduction of biological control agents - "Costa Lima" Quarantine Laboratory (CNPDA-EMBRAPA). Jaguariùna: EMBRAPA- CNPDA, 1995. 10 p.

FAO/IAE - USDA. Product quality control, irradiation and shipping procedures for mass-reared Tephritidae fruit flies for sterile insect realese programs. Vienna: IAEA, 1999, 51p.

FARIA, R. N. **Effects of the imposition of non-tariff barriers on Brazilian mango exports.** 2004, 127f. Dissertation (Master's Degree in Applied Economics) Federal University of Viçosa. Viçosa, MG, 2004.

FINNEY, G. L.; FISHER, T. W. Culture of entomophagous insects and their hosts. *In* **"Biological Control of Insect Pests and Weeds"** (P. De Bach, Ed.), Chapman & Hall, London. p. 328-353, 1964.

FOLLETT, P. A.; NEVEN, L. G. **Current trends in quarantine entomology.**

Annual Review of Entomology, Palo Alto, v. 51, p. 359-385, 2006.

FRANZ, G.; KERRREMANS, P.; RENDON, P.; HENDRICHS, J. Development and application of genetic sexing systems of the Mediterranean fruit fly based on a temperature sensitive lethal. In: MCPHERON, B. A.; STECK, G. J. **Fruit fly pests:** A world assessment on their biology and management. Delray Beach, Fl: St. Lucie Press, 1996. 185 p.

GALLO, D.; NAKANO, O.; CARVALHO, R. P. L; BAPTISTA, G. C.; BERTE FILHO, E.; PARRA, J. R.; ZUCCHI, R. A.; ALVES, S. B. VENDRAMIN, J. D.; MARCHINI, L. C.; LOPES, J. R. E.; OMOTO, C. **Entomologia Agricola**. 3. ed. Piracicaba: Fealq, 2002. 920 p.

GARCIA, M. A. Nutritional ecology of parasitoids and terrestrial predators. In:

PANIZZI, A. R.; PARRA, R. P. **Nutritional ecology of insects and its implications**

in pest management. Sâo Paulo, SP: Manole, 1991. chap. 8, p. 289-311.

HARAMOTO, F. H. **The biology of _Opius oophilus_ Fullaway (Braconidae - Hymenoptera)**. Master Thesis, University of Hawaii, USA. 1953.

HARRIS, E. J.; BAUTISTA, R. C. Effects of fruit fly host, fruit species, and host egg to female parasitoid ratios on the laboratory rearing of _Biosteres arisanus_.

Entomologia Experimentalis Et Applicata, v. 79, p. 187-194. 1996.

HEPDURGUN, B.;TURANLI, T.; ZÜMREOGLU, A. Parasitism rate and sex ratio of Psyttalia (=Opius) concolor (Hymenoptera: Braconidae) reared on irradiated Ceratitis capitata larvae (Diptera: Tephritidae). In: BLOEM, K.; GREANY, P.;

HENDRICHS, J. **Biocontrol Science and Technology**, v. 19, n. 1, p. 157-165, 2009.

HUNTER, M. K.; ROUX, E. A.; WOOD, R.J.; GILBURN, A. S. The effect of supra-fronto-orbital (sfo) bristle removal on male mating success in the Mediterranean fruit fly (Diptera: Tephritidae). **Florida Entomologist**, Gainesville, v. 85, n. 1,p. 83-88, 2002.

IBRAHM, A. G.; PALACIO , I. P.; ROHANI I. The life cycle of Biosteres arisanus , with reference to adult reproductive capacity on eggs of oriental fruit fly.

Malaysian Applied Biology, v. 21, p. 63-69. 1992.

BRAZILIAN FRUIT INSTITUTE (IBRAF). Available at:<http://www.ibge.gov.br/>. Accessed on: June 10, 2012.

KERREMANS, P.; FRANZ, G. Cytogenetic analysis of chromosome 5 from Mediterranean fruit fly, Ceratitis capitata. **Chromosoma**, Berlin, v. 103, p. 142146, 1994.

KNIPLING, E. F. Introduction. In **"Insect Colonization and Mass Production"** (C. N. Smith, Ed.), p. 1-12. Academic Press, New York. 1966.

KNIPLING, E. F. Possibilities of insect control or eradication through the use of sexually sterile males. **Journal of Economic Entomology**, College Park, v. 48, p.

459-462, 1955.

KOGAN, M. Integrated pest management historical perspectives and contemporary developments. **Annual Review of Entomology**, California, v. 43, n. 1, p. 243-270, 1998.

LASALLE, J.; GAULD, I. D. **Hymenoptera and Biodiversity**. C. A. B. International, 1993, 348p.

LAWRENCE, , P. O. The biochemical and physiological effects of insect hosts on the development and ecology of their insect parasites: an overview.

Archives of Insect Biochemistry and Physiology. Physiol, v. 13, p. 217-228, 1990.

LAWRENCE, P. O., BARANOSWSKI, R. M. GREANY, P. D. Effects of host age on development of *Biosteres (=Opius) longicaudatus*, a parasitoid of the caribbean fruit fly, *Anastrepha suspensa*. **The Florida Entomologist**, Gainesville, v. 59, n.1, p. 33-39, 1976.

LEA/ESALQ/USP. **Fruit flies (Diptera: Tephritidae) in Brazil**: hosts and parasitoids of the Mediterranean fruit fly Available at:

<http://www.lea.esalq.usp.br/ceratitis/edita_hosp_c.php>. Accessed on: July 5, 2014.

LINDQUIST, A. W. The use of gamma radiation for the control or eradication of the screwworn. **Journal of Economic Entomology**, Colleg Park, v. 48, p. 467469, 1955.

LIQUIDO, N. J. Effect of ripeness and location of papaya fruits on the parasitization rates of oriental fruit fly and melon fly (Diptera: Tephritidae) by braconid (Hymenoptera) parasitoids. **Environmental Entomology**, v. 20, p. 1732-1736. 1991

LOPES, A. P. S.; MORAES, M. C. B.; BORGES, M.; LAUMANN, R. A.

Oviposition behavior of the parasitoid Telenomus podisi (Hymenoptera: Scelionidae): influence of female age and host density on time allocation and parasitism ratio. In: XXII Congresso Brasileiro de Entomologia, 2008, Uberlândia, Science, technology and innovation: **Proceedings**.

Viçosa, MG: UFG.2008.

MALAVASI, A. Biology, life cycle, host relationship, important species and biogeography of tephritids. **In: International Training Course on Fruit Flies, 5.**, 2009, Vale do Sao Francisco, Brazil. Biology, monitoring and control of fruit flies. Juazeiro: Biofàbrica Moscamed Brasil, 2009. MALAVASI, A.; VIRGiNIO, J. F. (Ed.) p. 1-5.

MANOUKIS, N.; GEIB, S.; SEO, D. MCKENNEY, M.; VARGAS, R.; JANG, E. An optimized protocol for rearing *Fopius arisanus*, a parasitoid of tephritid fruit flies. Available at: http://www.jove.com/video/2901/an-optimized-protocol-for- rearing-fopius-arisanus-parasitoid. Accessed on: June 18, 2013.

MONTOYA, P., SUAREZ, A., LÓPEZ, F., CANCINO, J. *Fopius arisanus* **oviposition in four *Anastrepha* fruit fly species of economic importance in Mexico**. BioControl, v. 54,p. 437-444. 2009.

MORAES, G. J. de; SA, L.A.N. de; TAMBASCO, F. J. Brazilian legislation on the exchange of biological control agents. Jaguariùna: EMBRAPA - CNPMA, 1996, 16p. (EMBRAPA. **Documentos, 3**).

MORRISON, N. I.; SEGURA, D. F.; STAINTON, K. C.; FU, G.; DONNELLY, C.

A.; ALHEY, L. S. Sexual competitiveness of a transgenic sexing strain of the Mediterranean fruit fly, Ceratitis capitata. **Entomologia Experimentalis et Applicata**, Amsterdam, v. 133, p. 146-153, 2009.

MILLS, N. J. Parasitoid guilds: defining the structure the parasitoid communities endop insect hosts. **Environmental Entomology,** v. 23, n.5, p.1066-1083.

1994.

NAVARRO, M. A. **Trichogramma spp.: production, use and management in Colombia**. Guadalajara de Buga, Impretec Ltda., 176p. 1998.

NORRBOM, A.L. Updates to Biosystematic Database of World Diptera for Tephritidae through 1999. **Diptera Data Dissemination Disk** (CD-ROM) 2. 2004.

PAPADOPOULOS, N. T.; KATSOYANNOS, B. I.; CAREY, J. R. Demographic Parameters of the Mediterranean Fruit Fly (Diptera: Tephritidae) Reared in Apples. **Entomological Society of America**, v. 95, n. 5, p.564-569, 2002.

PARANHOS, B. A. J. Biological Control of Fruit Flies. In: MALAVASI, A.; VIRGINIO, J. (Ed.). **Biology, Monitoring and Control: V International Training Course on Fruit Flies.** Juazeiro, 2009. p. 2931.

PARANHOS, B. A. J. Sterile insect technique and biological control: environmentally safe and effective methods to combat fruit flies. In: SIMPOSIO DE MANGA DO VALE DO SAO FRANCISCO, 1., 2005, Juazeiro, BA.

Lectures... Petrolina: Embrapa Semi-Arido, 2005 (Documents, 189). Available

at: <http://www.agencia.cnptia.embrapa.br/recursos/OPB63ID- gwWrmJdPY.pdf>.

Accessed on: October 2, 2013.

PARANHOS, B. A. J.; NASCIMENTOS, A. S.; BARBOSA, F. R.; VIANA, R.;

SAMPAIO; MALAVASI, A.; WALDER, J. M. M. **Sterile Insect Technique**: new technology to combat the fruit fly, *Ceratitis capitata*, in the Sâo Francisco Valley. Petrolina: Embrapa Semiàrido, 2008. 6p. (Embrapa Semiàrido. Technical Communication, 137).

PARANHOS, B. A. J.; BARBOSA, F. R. Key pests in mango cultivation. In:

MACKAUER, M.; CHAU, A. Adaptive self superparasitism in a solitary parasitoid wasp: the influence of clutch size on offspring size. **Functional Ecology**, <u>v. 15, n. 3, </u>p.335-343, 2001.

PEDROSA-MACEDO, J. H. **Manual de Pragas em Florestas** - Forest Pests of Southern Brazil. IPEF/SIF, v. 2, 1993. 112p.

MENEZES, E. A.; BARBOSA, F. R. (Ed.). **Mango pests**: monitoring, level of action and control. Petrolina: Embrapa Semiàrido, 2005. chap. 2, p.51-69.

OKUNO, E.; YOSHIMURA, E. M. **Physics of radiation**. Sao Paulo: Oficina de Textos, 2010.

PARRA, J. R. P.; BOTELHO, P. S. M.; CORRÊA-FERREIRA, B. S.; BENTO, J.

M. S. Biological control: terminology. In: PARRA, J. R. P.; BOTELHO, P. S. M.;

CORRÊA-FERREIRA, B. S.; BENTO, J. M. S. (Ed.). **Biological control in Brazil: predatory parasitoids**. Sao Paulo: Manole, 2002. chap. 1, p. 1-16.

PEDIGO, L. P.; RICE, M. E. **Entomology and Pest Management**, Prentice Hall: New Jersey, 2005. 784 p.

POTENZA, M. R.; YASUO-KA, S.T.; GIORDANO, R. B. P.; RAGA, A. Irradiation of orange fruit infested with larvae of the fruit fly *Ceratitis capitata* (Wied., 1824). In: CONGRESSO BRASILEIRO DE ENTOMOLOGIA, 12., 1989. Belo

Horizon. **Abstracts...** Belo Horizonte: SEB, 1989. v.2, p.509.

PRATISSOLI, D.; OLIVEIRA, H. N. Influence of the age of Helicoverpa zea (Boddie) eggs on the parasitism of Trichogramma pretiosum Riley. **Pesquisa Agropecuària Brasileira**, v. 34, p. 891-896.

PURCELL , M. F. , JACKSON , C. G. , LONG , J. P. & BATCHELOR , M. A .Influence of guava ripening on parasitism of the Oriental fruit fly, *Bactrocera dorsalis* (Hendel) (Diptera, Tephritidae), by *Diachasmimorpha longicaudata* (Ashmead) (Hymenoptera, Braconidae) and other parasitoids. **Biological Control**, v. 4 , p. 396 - 403 . 1994

QUIMIO, G. M.; WALTER, G. H. Host preference and host suitability in an egg- pupal fruit fly parasitoid, *Fopius arisanus* (Sonan) (Hym., Braconidae). **Journal of Applied Entomology**. v. 125, p.135-140. 2001

RAGA, A.; YASUOKA, S. T.; AMORIM, E. O; SATO' M. E.; SUPLICY FILHO, N.; FARIA, J. T. de. Sensitivity of *Ceratitis capitata* (WIED., 1824) eggs irradiated in artificial diet and in mango fruits (*Mangifera indica* L.).

Scientia Agricola, Piracicaba, v. 53, n. 1, Apr./Jul., 1996.

RAGA, A.; SATO, M. E.; GIORDANO, R. B. P.; POTENZA, M. R.; SZULAK, C.; SUPLICY FILHO, N. Use of gamma radiation in the disinfestation of mangoes destined for export in relation to larvae of *Ceratitis capitata, Anastrepha fraterculus* and *Anastrepha obliqua.* In: BRAZILIAN CONGRESS OF ENTOMOLOGY, 13, 1991. Rio de Janeiro. **Abstracts...** Rio de Janeiro: SEB, 1991. v.2, p.632.

RAMADAN, M. M., WONG, T. T. Y., MCINNIS, D. O. Reproductive biology of Biosteres arisanus (Sonan), an egg-larval parasitoid of the oriental fruit fly. **Biological Control** , v. 4, p. 93 -100, 1994

RAMADAN, M. M.; WONG,T. T. Y.; BEARDSLEY, J. W. Reproductive behavior of *Biosteres arisanus* (Sonan) (Hymenoptera: Braconidae), an egg-larval parasitoid of the oriental fruit fly. **Biological Control,** v. 2, p. 28-34, 1992.

RENWICK, J. A. A. Chemical ecology of oviposition in phytophagous insects. **Experientia,** v. 45, p. 223-228, 1989.

ROBINSON, A. S.; FRANZ, G.; FISHER, K. Genetic sexing strains in the medfly, Ceratitis capitata: development, mass rearing and field application. **Trends in Entomology**, Kerala, v. 2, p. 81-104, 1999.

ROCHA, K. L.; MANGINE, T.; HARRIS, E. J.; LAWRENCE, P. O. Immature Stages of *Fopius arisanus* (Hymenoptera: Braconidae) in *Bactrocera Dorsalis* (*Diptera: Tephritidae*). **Florida Entomological Society,** v. 87, n. 2, p. 164-168, 2004.

ROITBERG, B. D.; BOIVIN, G.; VET, L. Fitness, parasitoids and biological control: an opinion. **Canadian Entomologist**, 133: p. 429-438. 2001.

ROSSLER, Y The genetics of the Mediterranean fruit fly: a "white-pupa" mutant. **Annals of the Entomological Society of America**, Lanham, v. 72, p. 583-590. 1979a

ROSSLER, Y Automated sexing of Ceratitis capitata (Dip. Tephritidae): The deveplopment of strain with inherited sex-limited pupae color dimorphism.

Entomophaga, Paris, v. 24, p. 441['. 1979b.

ROUSSE, P., HARRIS, E. J., QUILICI, S. *Fopius arisanus*, an egg-pupal parasitoid of Tephritidae. Overview. **Biocontrol News and Information** v. 26, n. 2, p. 59-69, 2005.

SA, L. A. N. Quarantine and the exchange of biological control agents. **O Biològico,** Sao Paulo, v.2, n. 2, p.1-6, 2001.

SA, L. A. N. de; TAMBASCO, F. J.; LUCCHINI, F. Import, export and regulation of biological control agents in Brazil. In: BUENO, V. H. P. (Coord.). **Quality control of biological control agents**. Lavras: UFLA, p. 187-196, 1999.

SEGURA, D. F.; VISCARRET, M. M.; OVRUSKI, S. M.; CLADERA, J. L. Response of the fruit fly parasitoid Diachasmimorpha longicaudata to host and host-habitat volatile cues. **Entomologia Experimentalis et Applicata**, v. 143, p. 164-176, 2012.

SEQUEIRA, R.; MACKAUUER, M. **Seasonal variation in body size and offspring sex ratio in field populations of the parasitoid wasp, *Aphidius ervi* (Hymenoptera: Aphidiidae)**. Oikos. v. 68, p. 340-346. 1993.

SILVA, J. W. P.; BENTO, J. M. S.; ZUCCH, R. A. Olfactory response of three parasitoid species (Hymenoptera: Braconidae) to volatiles of guavas infested or not with fruit fly larvae (Diptera: Tephritidae). **Biological Control**, v. 41, p. 304311, 2007.

SILVA NETO, A. M.; SANTOS, T. R. O.; DIAS, V. S.; JOAQUIM-BRAVO, I. S.; BENEVIDES, L. J.; BENEVIDES, C. M. J.; SILVA, M. V. L.; SANTOS, D, C. C.; VIRGINIO, J.OLIVEIRA, G. B.; WALDER, J. M. M.; PARANHOS, B. A. J.; NASCIMENTO, A. S. Mass-rearing of Meditarranean fruit fly using low-cost yeast products produced in Brazil. **Scientia Agricola**, v. 69, n. 6, p. 364-369. 2012.

SINGER, M. C. The definition and measurement of oviposition preference in plant-feeding insects. In: MILLER, J. R.; MILLER, T. A. (Ed). **Insect-Plant Interactions**. Springer-Verlag, New York, p. 65-94, 1986.

SIVINSK, J.; SMITTLE, B. Effects of gamma radiation on the development of the Caribbean fruit fly (Anastrepha suspensa) and the subsequent development of its parasite Diachasmimorpha longicaudata. **Entomologia Experimentalis et Aplicata**, Dordrecht, v. 55, p. 295-297, 1990.

SNOWBALL, G. J.; WILSON, F.; CAMPBELL; T. G. AND LUKINS, R. G. The utilization of parasites of oriental fruit fly (Dacus dorsalis) against Queensland fruit fly (Strumeta tryoni). **Australian Journal of Agricultural Research**, v. 13, p. 443-460, 1962

STEFANELO, E. L. Agronegócio Brasileiro: Propostas e tendências. **FAE BUSINESS**, Curitiba, n. 3, p. 10-13, 2002.

STUHL C.; SIVINSKI J.; TEAL P.; ALUJA, M. Responses of multiple species of Tephritid (diptera) fruit fly parasitoids (Hymenoptera: braconidae) to sympatric and exotic fruit volatiles. Gainesville, **Florida Entomologist**, v. 95, n. 4, p. 1031-1039, 2012.

VAN ALPHEN, J. J. M.; JERVIS, M. Foraging behavior. In: JERVIS, M.; KIDD, N. (Eds.) **Insect natural enemies**. London: Chapman and Hall, 1996. p, 1-62.

VARGAS, R. I., PECK, S. L., McQUATE, G. T., JACKSON, C. G., STARK, J. D.; MRMSTRONG, J. W. Potential for areawide integrated pest management of Mediterranean fruit fly (Diptera: Tephritidae) with a braconid parasitoid and a novel bait spray. **Journal of Economic Entomology**. v. 94, n. 4, p. 817-825. 2001.

VARGAS, R. I. , STARK, J. D. , PROKOPY, R. J.; GREEN, T. A. Response of oriental fruit fly (Diptera: Tephritidae) and associated parasitoids (Hymenoptera: Braconidae) to different-color spheres. **Journal of Economic Entomology**, v. 84 , p. 1503 - 1507. 1991.

VINSON, S. B.; IWANTSCH, G. F. Host Sustability for insect parasitoids.

Annual Review of Entomology. v. 25, p. 397-419. 1980.

WALDER, J. M. M. Sterile Insect Technique - Genetic Control. In: MALAVASI, A.;

ZUCCHI, R. A. (Ed.). **Fruit flies of economic importance in Brazil:**

Basic and applied knowledge. Ribeirao Preto: Holos, 2000. chap. 19, p. 151- 158.

WALDER, J. M .M.; COSTA, M. L. Z. Gamma radiation as a disruptive agent of the defense mechanism of *Ceratitis capitata* (Wied. 1824) (Diptera: Tephrididae) larvae against *Diachasmimorpha longicaudata* (Ashmead) (Hymenoptera: Braconidae). IN: SIMPÒSIO DE CONTROLE BIOLÒGICO, 8. 2003, Sâo Pedro. **Abstracts...**Sâo Pedro, 2003. p.138.

WALDER, J. M. M.; MASTRAGELO, T. A. X-rays: new technology for the sterile insect technique. In: 12th SICONBIOL, Symposium on Biological Control: "Climate Change and Sustainability: Breaking Paradigms". 2011.

WANG, X; MESSING, R. H. Foraging Behavior and Patch Time Allocation by *Fopius arisanus* (Hymenoptera: Braconidae), an Egg-Larval Parasitoid of *Tephritid* Fruit Flies **Journal of Insect Behavior**, v. 16, n. 5, p. 593-612, 2003.

WHARTON, R. A.; F. E. GILSTRAP. Key to and status of opiinae braconid (Hymenoptera) parasitoids used in biological control of *Ceratitis capitata* and *Dacus* s.l. (Diptera: Tephritidae). **Annals of the Entomological Society of America.** v. 76: p. 721-742, 1983.

ZANARDI,O. Z.; NAVA, D. E.; BOTTON, M.; GRÜTZMACHER, E. D.; MACHOTA JR, R.; BISOGNIN, M. Development and reproduction of the Mediterranean fruit fly in coconut, apple, peach and grapevine trees. **Pesquisa Agropecuària Brasileira**, Brasilia, v.46, n.7, p. 682-688, 2011.

ZENIL, M.; LIEDO, P.; WILLIAMS, T.; VALLE, J.; CANCINO, J.; MONTOY, P. Reproductive biology of *Fopius arisanus* (*Hymenoptera: Braconidae*) on *Ceratitis capitata* and *Anastrepha spp.* (*Diptera: Tephritidae*). **Biological Control**, Chiapas, v. 29, p. 169-178, 2004.

ZUCCHI, R. A. Diversity, distribution and hosts of the genus Anastrepha in Brazil. In: HERNANDEZ-ORTIZ, V (Ed.). **Fruit flies in Latin America (Diptera: Tephritidae):** diversity, biology and management. Mexico, S y G Editores, 2007.

p.77-100.

ZUCCHI, R. A. Mediterranean fruit fly, *Ceratitis capitata* (Diptera: Tephritidae). In: VILELA, E. F.; ZUCCHI, R. A.; CANTOR, F. (Ed.). **History and impact of introduced pests in Brazil.** Ribeirâo Preto: Holos, 2001. chap. 1, p. 15-22.

ZUCCHI, R. A. Taxonomy. In: MALAVASI, A.; ZUCCHI, R. A. (Eds.). **Fruit flies of economic importance in Brazil.** Basic and applied knowledge. Ribeirao Preto: Holos, 2000. p.13-24.

yes
I want morebooks!

Buy your books fast and straightforward online - at one of world's fastest growing online book stores! Environmentally sound due to Print-on-Demand technologies.

Buy your books online at
www.morebooks.shop

Kaufen Sie Ihre Bücher schnell und unkompliziert online – auf einer der am schnellsten wachsenden Buchhandelsplattformen weltweit! Dank Print-On-Demand umwelt- und ressourcenschonend produziert.

Bücher schneller online kaufen
www.morebooks.shop

Printed by Books on Demand GmbH, Norderstedt / Germany